Teaching & Education Without Burnout

How AI Tools Empower Teachers, Future-Proof Students, Enable Innovation, Unlock Personalized Learning, Save Time, Improve Outcomes, & Reignite A Love of Learning

Keryn Gold, PhD, MBA

Lead With Gold

Author: Keryn Gold, PhD, MBA
Publisher: Lead With Gold, LLC

Legal Notice

Paperback ISBN: 978-1-967456-16-1
Hardcover ISBN: 978-1-967456-17-8
eBook ISBN: 978-1-967456-15-4
Printed in the United States of America
First Edition, 2025
LeadWithGold.com

Contents

To some, the phrase "AI in the classroom" raises more concerned questions than answers: Will it replace teachers? Will students become passive or cheat their way through? We're here to rewrite that narrative. In reality, well-designed AI tools free educators from repetitive tasks—like constant grading and prep—so you have space to do the soul and mission-driven work that inspired you to teach in the first place. It's about focusing on empathy, creative problem-solving, and big-picture lessons while leveraging AI's efficiency to tailor learning experiences like never before.

And yes, every adult can be part of this revolution—because education has never solely belonged within four walls. Parents help shape their children's learning journeys at home, community leaders champion innovative programs, and students themselves become co-creators in their own growth. For educators, the question is simple yet profound: **How do we harness AI to build up our potential as leaders, to free ourselves from burnout, and to raise a generation that's entrepreneurial, empathic, unafraid to experiment, and empowered to thrive in the world of tomorrow?**

Let's blend the rigor of academic research with the heart of real-life teaching. Let's apply behavioral insights to help you—and your students—push beyond fear and limitation. Let's unify around a central purpose: to educate in ways that spark lasting curiosity, moral courage, and readiness for whatever tomorrow's AI-charged world may bring.

In these pages, you'll find practical strategies that show how AI can customize lesson plans for different skill levels, lighten your grading load, and amplify your ability to personally connect with students. You'll see how to foster creativity, entrepreneurship, AI and code literacy, leadership, and emotional intelligence—skills that an algorithm alone can't teach but

Introduction: Standing at the Crossroads of Frustration and Opportunity

—Break Free From Burnout & A Broken System With the Help of Generative AI

Take a moment to envision a teaching world where burnout no longer consumes our brightest educators. Picture a classroom humming with curiosity and creativity, where lessons adapt to each learner at just the right time. Imagine feeling *less* like an overworked manager of endless tasks and *more* like a mentor shaping bold, future-ready minds. That's the promise of generative AI in education—a promise we can fulfill right now by uniting our passion for teaching with transformative technology.

can *support* you in teaching better. And you'll discover a framework that respects your humanity and your students' unique paths, all while charting out how to use AI responsibly and ethically in the classroom and at home.

This isn't about technology overshadowing teachers; it's about **teacher empowerment**. It's about forging forward as educators who innovate, who adapt, who push each student to become not just test-takers or data points, but confident, compassionate, and entrepreneurial citizens of a dynamic world. Most importantly, it's about reclaiming the joy and meaning in teaching—because when we liberate ourselves from burnout, we light the spark of lifelong wonder in our students, fueling a brighter, more collaborative future for all.

Welcome to the next chapter in education—where your leadership meets cutting-edge AI, and where your passion for teaching turns into unstoppable impact without the weight of daily grind. Let's ignite our collective vision for schools, families, and kids who are prepared—and *excited*—for an extraordinary tomorrow.

1

Personalized, Not Overwhelmed: How AI Frees Teachers from Burnout and Meets Every Student Where They Are

— Give each learner a custom roadmap – while lightening your load and elevating your impact.

Picture walking into class each morning knowing every student—whether eager overachiever or reluctant learner—has exactly the right challenge waiting for them. Imagine hitting your midweek stride without feeling drained by endless grading and always feeling like you need to "play catch-up." That vision becomes real when we bring AI-powered personalization into the classroom. Not as a tech gimmick or a one-off experiment, but as a real strategy for liberating educators from repetitive tasks and giving them the freedom to focus on deeper, more human-centered teaching.

From One-Size-Fits-All to True Individualization

For generations, teachers have been stuck managing a standardized curriculum that marches all students forward in lockstep. The result? Some learners fall behind, others get bored, and teachers spin their wheels trying to differentiate for 30 kids simultaneously, whose capabilities may span 5 or more grade levels, in the same classroom. No wonder so many of us end the year feeling exhausted, under-resourced, and—frankly—burnt out.

Enter AI-driven personalization. Rather than having to assign the same math worksheet or reading chapter to everyone, you can now let technology parse out each student's exact skill level—filling gaps, accelerating advanced learners, and offering a steady flow of right-sized challenges. When done appropriately, this shift gives you immediate data on who needs help, who's ready to move on, and how best to channel your energy.

Key Benefit 1: Liberate Time and Energy

- **Automate Repetitive Tasks**
  AI can grade quizzes, assign adaptive practice, and track progress in real time. You save hours of manual tallying, allowing you to coach and connect with students more effectively, and with more energy.

- **Instant, Actionable Data**

Instead of waiting for a test to reveal which students are missing crucial skills, AI updates you daily—or even hourly—enabling you to make quick adjustments that preempt burnout and disengagement, and prevent student meltdowns.

Key Benefit 2: Turbocharge Student Engagement

- **Right Challenge, Right Moment**
  Students rarely get bored or lost because lessons adapt to their pace—like having a personal tutor who never judges and never tires.

- **Immediate Feedback Loops**
  Learners see corrections on the spot, building confidence and turning mistakes into stepping stones instead of frustrating struggles. There's a clear path forward, and clear guidance, at the right level for them.

Key Benefit 3: Restore A Genuine Love of Teaching & Learning

- **Room for Human Connection**
  By offloading routine drill practice to AI, you can use freed-up time for creative lessons, group discussions, or one-on-one emotional support and coaching, to truly connect with your students.

- **Reduced Stress & Burnout**
  When "the grind" no longer swallows your schedule, you can focus on leading and innovating and empowering—reclaiming the reason you became an educator in the first place.

How AI Personalization Works

A good AI platform continuously evaluates each student's performance, adjusting content or difficulty on the fly. Think "GPS for learning." Instead of giving everyone identical arithmetic drills, the system delivers practice sets targeted at each learner's demonstrated needs. That means no more "teaching to the middle" or re-explaining the same fractions over and over. The moment the AI detects improvement, it nudges the student upward. If a concept stumps them, it refines the approach to shore up gaps in understanding.

Real-World Example

Khan Academy AI & DreamBox:

1. **Assessment** – The program identifies where the learner struggles or excels.
2. **Adaptive Assignments** – It assigns custom tasks aligned with those findings.
3. **Instant Feedback** – Students see hints and corrections immediately.
4. **Progress Overview** – Teachers view a dashboard highlighting which students need more personal intervention vs. who's sailing ahead.

The result? Each student's path is constantly in sync with their growth curve, requiring minimal teacher micromanagement.

Practical Tips for Teachers to Start Personalizing

1. **Pilot in One Subject**
   Instead of overhauling your entire curriculum, pick one area—say 7th-grade math or reading comprehension—and integrate an AI platform there. Track changes in student engagement and test scores.

2. **Mix AI with Humanity**
   AI handles practice drills or initial feedback. Meanwhile, you run deeper dialogues, small-group sessions, and creative projects. Students still benefit from your expertise, and you get to do the "fun" parts more often.

3. **Encourage Student Ownership**
   Teach kids to read their AI-based progress reports and set personal goals. Encourage them to reflect on how technology is enhancing their learning rather than just passively following prompts.

4. **Stay Flexible & Monitor Progress**
   AI can highlight issues or recommend next steps, but you, as the educator, remain the final authority. Trust your instincts—if a student needs to slow down or approach a topic differently, override the system as needed.

Ethical and Responsible Implementation

While AI personalization opens incredible possibilities, keep guardrails in place:

- **Data Privacy**: Ensure platforms comply with FERPA, COPPA, and similar regulations.

- **Bias & Transparency**: Confirm that the algorithms, or the datasets they rely on, don't disadvantage certain groups or stifle diverse perspectives.

- **AI Literacy**: Teach students (and fellow staff) how to question AI outcomes. Show them that even a sophisticated algorithm can make flawed assumptions. Ensure appropriate guardrails are in place, and teachers and students know how to use AI appropriately, as an empowerment tool.

Overcoming Common Objections

- **"Won't the computer do all the teaching?"**
  Absolutely not. Teachers remain indispensable for relational support, advanced discussions, and orchestrating real-world applications. AI is an assistant, not a substitute for human empathy and creativity.

- **"My class is too chaotic for new tech."**
  Start modestly—introduce AI practice in short, controlled segments. As students see improvements (and you feel less stressed), buy-in will grow organically.

- **"Not enough devices or funding."**
  Look for grant opportunities, pilot programs, or open-source AI solutions. Even limited device usage can make a significant impact when used efficiently (e.g., rotating stations). We'll discuss strate-

gies for this in depth in Chapter 11.

Conclusion: Personalization as a Path to Empowerment

When the daily grind no longer holds you hostage, and each student's needs no longer overwhelm your planning, the entire teaching experience transforms. AI personalization **amplifies** your reach, **deepens** your connection to learners, and **relinquishes** the one-size-fits-all script that has fueled so much burnout.

This isn't about displacing the magical spark that only real educators can ignite. It's about freeing that spark to shine brighter. As you integrate AI-driven individualization, you'll see that students flourish when tasks match their readiness—and you'll flourish too, unburdened by repetitive tasks, and able to breathe more life, creativity, and purpose into your classroom.

"Technology by itself isn't the solution. It's a catalyst that lets us focus on the true art of teaching: guiding minds, building confidence, and shaping ethical, entrepreneurial humans who can adapt to a world brimming with possibility."

Next Up: In **Chapter 2**, we'll tackle the biggest fears and myths swirling around AI in the classroom—like the idea that it replaces teachers or encourages lazy students—and show you how to cut through the noise to make informed, empowering decisions. Your role as an educator only gains strength as you discover how to dispel these misconceptions and lead with clarity, trust, and innovation. Let's get started.

2

Debunking AI Myths: Empower Teachers, Don't Replace Them

— Break free from fear and misconceptions—embrace AI as a supportive ally that magnifies your influence

There's a persistent buzz in education circles that if we invite AI into our classrooms, we're somehow diminishing the teacher's role—perhaps even setting ourselves up for obsolescence. Others worry AI will produce lazy learners who let algorithms do the heavy lifting instead of developing true critical thinking skills. It's time to clear the air. AI, when used thoughtfully, **expands** what teachers can accomplish and **motivates** students to **engage more deeply**. This chapter tackles the most pervasive myths around AI in schools, shedding light on how we can bust these misconceptions and harness its real potential.

Myth #1: "AI Will Replace Teachers"

Reality: AI Eases Your Workload; It Doesn't Substitute Your Humanity

Despite Hollywood visions of robots taking over, no algorithm can replicate the empathy, leadership, and creativity you bring to the classroom. AI excels at automating tasks (like grading repetitive worksheets) or delivering targeted practice. You remain indispensable for the intangible elements—coaching, mentoring, building community, and fueling a lasting sense of wonder.

- **What AI Can Do**
    - Adapt lessons to each student's pace.
    - Provide immediate feedback on drills or quizzes.
    - Highlight gaps for you to address in small-group or targeted lessons.
- **What Only Teachers Can Do**
    - Form emotional bonds, sense nonverbal cues, and respond to deeper personal needs.
    - Spark curiosity, pose big questions, and help students connect learning to real-life challenges.
    - Serve as moral and intellectual role models, guiding students on how to learn.

Key Point:

AI is more like a flexible assistant than a teacher replacement—freeing you from mundane tasks so you can step up as the creative, empathic leader your classroom craves.

Myth #2: "AI Makes Students Lazy and Dependent"

Reality: AI Tools Ignite Curiosity and Accelerate Mastery, If Used Correctly

Yes, if students simply copy-paste AI-generated text without reflection, they're not learning. But that's like blaming a calculator for poor math skills—misuse isn't the tool's fault. Properly guided AI actually cultivates **self-directed learning** and **deeper engagement** by removing the guesswork and frustration from routine tasks.

- **Responsible Classroom Practices**
  - **Set Clear Guidelines**: Have students show their steps or discuss how they arrived at AI-aided solutions.
  - **Promote Verification**: Encourage students to compare AI outputs with real-world examples or multiple sources.
  - **Spark Creative Extensions**: AI might generate basic outlines or initial ideas; students then add unique insights, research, or personal flair. Teach them how to use AI correctly, and you set them up for success in the future. Over 50% of online content is already AI-generated, and by the end of this year,

over 90% of code will be written by AI. This is staggering. We do our students a disservice by not teaching them to use AI responsibly.

Key Point:

Students become lazier only if we allow AI to serve as a digital shortcut. The real opportunity is to leverage AI for brainstorming, skill-building, and iterative feedback—while making sure kids still do the "heavy lifting" of thinking and understanding AI's insights.

Myth #3: "AI is Biased and Untrustworthy—We Can't Trust Its Content"

Reality: AI Reflects the Data It's Trained On; Teach Students to Recognize Bias

Bias can absolutely exist in AI models; they inherit the strengths and flaws of their training data. But that's not a reason to discard AI altogether. Instead, we should **empower students** (and ourselves) to understand how AI generates answers, question potential bias, and fact-check the outputs.

- **How to Address AI Bias**
  - **Explainable AI**: Choose tools that share how they arrive at conclusions (or let students see the steps). Perplexity AI is a great example of this, and an AI-powered search engine that explicitly cites all sources.
  - **Critical Literacy**: Teach kids how to cross-reference AI re-

sponses, just as they would with any source.

- **Diverse Training Samples**: When selecting AI platforms, look for those aiming to reduce bias by training on varied datasets.

Key Point:

AI bias can become an incredible teachable moment: we show kids how to critique information sources—be they AI or human—to build informed skepticism and ethical responsibility.

Myth #4: "We Don't Have Enough Funding, Devices, or Bandwidth for AI"

Reality: Start Small, Seek Grants, and Pilot Strategically

Not every district can afford a high-tech overhaul. But AI integration doesn't require a lab of cutting-edge devices. Many AI platforms run effectively on standard computers or tablets, and some offer free versions to schools.

- **Practical Tips**
  - **Pilot in One Grade or Subject**: Let's say you focus on 6th-grade math with Khan Academy AI, track results, build a success story, and then seek grants or sponsors to expand.
  - **Rotate Device Access**: If you have limited hardware, use a station-rotation model so students share AI practice time.

- **Open-Source/Low-Cost AI**: Tools like Koala or certain no-frills adaptives can fit tight budgets, especially for initial trials. Contingent, risk-free cost savings consultancies can also find money in your existing budget, without hiring freezes, layoffs, or reductions in resources or quality. They get paid out of the savings they find, so there is no net cost. This and other funding sources are address in more detail in Chapter 11.

Key Point:

Implementation can be scaled—embrace incremental progress, gather data, and demonstrate early wins to secure broader funding or community support.

Myth #5: "Our Teachers Aren't Tech-Savvy Enough for AI"

Reality: Modern AI Tools Are More User-Friendly Than You Think

Many AI platforms feature intuitive dashboards and minimal setup. With a bit of training—and a willingness to explore—teachers quickly discover that AI reduces stress, rather than adding complexity. Districts can offer in-house training or connect teachers to simple online tutorials.

- **Professional Development Approach**
    - **Hands-On Workshops**: Let teachers practice AI tasks for lesson planning, grading, or generating interactive practice sets.

- **Mentor-Teacher Model**: Identify one or two "tech enthusiasts" who can troubleshoot and guide others.

Key Point:

If you're comfortable with email or a smartphone, you can likely navigate an AI tutoring tool. The real leap is mental—overcoming the fear that "this is too advanced" and seeing how seamlessly AI fits everyday teaching. Model the same growth mindset you seek to foster in your students.

Action Steps to Bust Myths in Your School

1. **Start Myth-Busting Conversations**
    - Host an informal staff meeting or parent workshop. Show real examples of how AI supports (not supplants) teacher autonomy.
2. **Deploy a Small Pilot**
    - Prove that AI simplifies tasks—like essay feedback or practice drills—without harming academic integrity or creativity. Share success stories to reduce skepticism.
3. **Invite Skeptics to Observe**
    - Let reluctant teachers see a colleague's class that uses AI effectively. Seeing is believing.
4. **Encourage Student Reflections**
    - Students who've experienced AI's benefits can speak to im-

proved engagement, less frustration, and excitement over personalized challenges that nurture a true love of learning.

Key Takeaways

1. **AI Never Replaces Human Teachers**
    - It eliminates mundane tasks, allowing you to focus on genuine mentorship, empathy, and vision.
2. **Students Aren't Lazy if AI Is Used Properly**
    - Responsible guidelines + critical reflection = deeper learning and skill mastery.
3. **Bias in AI Exists—but It's a Learning Opportunity**
    - Teach kids to question data sources, cross-check info, and become ethical digital citizens.
4. **Budget & Tech Hurdles Are Surmountable**
    - Start small, test and refine, then leverage success stories for grants or broader rollouts.
5. **Teachers Can Master AI, Even If They're Not 'Tech Gurus'**
    - Most modern platforms are user-friendly; a little professional development goes a long way.

By challenging these myths head-on, you'll feel more confident in introducing AI and guiding your school community toward beneficial, future-driven learning.

Next Up: In **Chapter 3**, we explore the **Future-Ready Skillset**—helping your students thrive in an AI-driven world by cultivating creativity, entrepreneurial thinking, and emotional intelligence. Get ready to see how AI can unlock skill sets that standardized curricula often overlook, making you and your students unstoppable in the face of constant change.

3

Future-Proof Skills: Equipping Students with Creativity, AI Literacy, and Entrepreneurial Mindsets

– Spark the curiosity, adaptability, and leadership your learners need to thrive in an AI-driven era.

Each day, technology nudges the world forward at a pace that can feel dizzying—even overwhelming. Gone are the times when rote memorization and standardized exam performance could guarantee a stable career. Today's students need to become *agile, inventive, and fearless* in how they tackle new challenges. They must learn how to collaborate with

AI rather than compete against it or ignore it, and how to harness new opportunities with an entrepreneurial lens.

This chapter uncovers **the top skills** your students (and you) can cultivate to thrive and avoid falling behind in an AI-driven future. You'll see how to marry AI-based learning with real-world problem-solving, ensuring that your learners grow into **critical thinkers, creative explorers, ethical innovators, and confident leaders**.

Why Traditional Education Falls Short

Outdated benchmarks—like purely test-focused instruction—reward memorization over genuine understanding. Meanwhile, skill sets like digital fluency, creativity, emotional intelligence, and entrepreneurial thinking often get sidelined as "extras." But the roles that rely solely on memorized content or routine tasks are the very ones AI can easily replicate. So the solution is to push beyond memorization and standardized paths, **letting students explore, build, and lead** in ways technology alone can't.

Essential Skills for an AI-Powered World

1. Critical Thinking & Problem-Solving

Why It Matters: AI is great at pattern recognition and data analysis, but it can't replicate human insight into context, bias, or moral complexity.

- **Real-World Example**: A high school history class uses ChatGPT to generate a summary of a historical event, then analyzes possible biases. Students dissect *why* the AI might have omitted certain details, strengthening their research acumen.

- **Implementation Tip**: Encourage students to pose "What if...?" or "Why not...?" questions about AI-generated answers, fostering an active rather than passive approach.

2. Digital & AI Literacy

Why It Matters: Just as reading, writing, and arithmetic were essential a century ago, understanding how AI tools work and when to trust (or question) them is crucial now.

- **Real-World Example**: A middle school science class uses an AI platform to interpret lab data. Students must discuss potential flaws in the dataset—learning how AI can misread incomplete or skewed inputs.

- **Implementation Tip**: Add short, frequent modules about how algorithms gather, process, and interpret data. This mini "AI 101" ensures every learner knows the basics of what's happening under the hood.

3. Creativity & Innovation

Why It Matters: AI can generate text or designs quickly, but humans excel in weaving imagination, empathy, and personal vision into solutions.

- **Real-World Example**: A classroom brainstorming session merges a platform like DALL·E, Canva, or Ideogram for visual prototypes with hands-on crafts. Students produce AI-inspired design sketches, then iterate physically to refine or deviate from the initial concept.

- **Implementation Tip**: Regularly run "AI + Real-World" challenges, letting students blend algorithmic sparks with their own artistry—be it storytelling, music composition, or product prototyping.

4. Entrepreneurial & Financial Literacy

Why It Matters: As AI disrupts traditional career paths, those who can spot opportunities, manage resources, and pivot quickly remain unstoppable.

- **Real-World Example**: An 8th-grade project tasks learners with forming small businesses. They use AI-driven market research tools (like Perplexity or ChatGPT) to identify trends, then design prototypes they pitch to peers. They can do this, today, for free.
- **Implementation Tip**: Have students keep digital "business journals" about cost calculations, marketing copy, and user feedback—backed by AI analytics. This merges math, writing, and problem-solving skills with a real entrepreneurial spark.

5. Emotional Intelligence & Adaptability

Why It Matters: AI is logical; humans must excel at empathy, leadership, and mindful collaboration. Being flexible means thriving in roles that shift as tech evolves.

- **Real-World Example**: Students practice "collaborative coding," where one writes code with Replit while others advise on design or user experience. The group must handle disagreements or unexpected AI code suggestions gracefully.

- **Implementation Tip**: Infuse structured reflection times—where students record what emotions came up during an AI-led task, how they handled them, and how they adapted or compromised. Emotions become teachable data and invaluable life experience relevant to real-world problem-solving and dispute resolution.

How Educators Can Foster These Skills

1. **Embrace Project-Based Learning**
   - Let students pick real-world issues—climate, local community challenges—and partner with AI tools for brainstorming or data analysis. They'll gain a sense of ownership and see how to direct technology productively.
2. **Teach Students to Evaluate AI Outputs**
   - From fact-checking claims to analyzing potential biases, embed a "trust but verify" mindset. This transforms AI from an authority to a *collaborator* they critically engage with.
3. **Incorporate Light Entrepreneurship**
   - Even if you're not a business teacher, short sessions on idea and concept generation, validation, market and competitive analysis, budgeting, or marketing can be applied to any subject. AI's capacity to do quick research or generate pitch scripts fosters mini ventures in the classroom that students find incredibly

engaging.

4. **Schedule Reflection & Group Dialogue**
   - After AI-based tasks, have students (and yourself) ask: "What did I learn beyond the content? How did the tool's suggestions change my thinking? What would I do differently next time?" This cements both emotional intelligence and adaptability.

Success Story Snapshot

Case: A school in an economically challenged area integrates an AI-based entrepreneurship module for 9th graders. Students use free versions of AI marketing tools to study local needs, design simple products (handmade items, digital goods), and pitch them at a student fair.

- **Outcome**: Over 60% of participants mention improved confidence in problem-solving and basic financial planning, plus excitement about coding tie-ins for potential e-commerce. Teachers see a drop in disengagement—students love the real-world flair and quick AI feedback loops.

Key Takeaways

1. **Memorization Isn't Enough**: To thrive in the AI era, learners require critical thinking, digital/AI literacy, creativity, entrepreneurship, and emotional intelligence.

2. **AI as an Accelerator**: Tools assist in coding, brainstorming, or analysis, but real learning requires active student engagement and

teacher-led context.

3. **Real-World Relevance**: Encourage ventures, creativity, collaboration—key qualities that no algorithm can replicate.
4. **Reflection & Adaptation**: Shift from "did you get the right answer?" to "how did you approach and refine your ideas?"—fueling deeper personal growth.

Next Up: In **Chapter 4**, we'll dive into "AI as a 1:1 Tutor—Practical Strategies for Educators." You'll see how adaptive learning tools transform skills practice for your students, giving you more bandwidth to deliver rich connections, advanced problem-solving, and the vital guidance that only a caring human can provide.

4

AI as Your 1:1 Tutoring Force: Practical Strategies to Elevate Every Classroom

– Offer personalized, data-backed support—then guide, mentor, and inspire in ways only you can

Imagine each of your students having instant, on-demand 1:1 help—like a private tutor but with infinite patience, real-time diagnostics, and zero judgment. That's the promise of AI-powered tutoring. While AI handles repetitive drills, immediate corrections, and customized pacing, **you** remain the mentor who coaches deeper understanding, empathy, and real-world application. This chapter lays out how to integrate AI tutoring into everyday practice, creating a synergy that energizes you and boosts your students' mastery.

Why AI Tutors Are Game-Changers

Overcoming "One-Size-Fits-All" Limits

Traditional instruction can unintentionally leave some students behind while others coast along unchallenged, becoming disengaged. With **AI-driven** adaptive tutoring, learners advance at the speed right for them—never bored, never lost. You finally escape the logistical nightmare of trying to differentiate manually for each student in a class of 30 with skill levels for each subject spanning 5+ grade levels.

- **Instant Feedback**: AI spots errors the moment they happen, suggesting hints or simpler tasks if a concept's not sticking.
- **Real-Time Data**: You see who's struggling with fractions, who's acing algebra, and who needs more creative challenges.

Educator Insight:

It's like having a volunteer tutor for each kid—answering queries, repeating explanations if needed, and never running out of energy.

How AI Tutors Work in Practice

Step 1: Identify the Right Platform

Several AI tutoring tools exist; pick one that aligns with your subject, age group, and budget. Examples include:

- **Khan Academy AI**: Broad coverage (math, science, humanities), free to most.
- **DreamBox**: Deep focus on K–8 math, adjusting path on the fly.
- **Photomath**: Great for step-by-step math solutions and immediate clarifications.

Tip: Start small—pilot in one class or one skill area. Evaluate student engagement and progress data before expanding.

Step 2: Implement in Your Classroom Routine

- **Rotation Model**: Students spend 10–15 minutes on the AI tool, freeing you for small-group or one-on-one sessions.
- **Homework/Flipped Classroom**: Assign AI-based drills or reading at home, saving class time for interactive exploration.
- **Station Collaboration**: Some students work with AI individually; others do group tasks. Rotate so every learner gets personalized practice.

Step 3: Oversee Data, Then Intervene

AI tools typically provide dashboards or quick snapshots of each student's performance:

- **Highlight Gaps**: If Sarah repeatedly struggles with exponents, you schedule a mini-lesson or a quick check-in.
- **Track Growth**: If Devon leaps two grade levels in a month, you applaud them and provide advanced tasks or creative projects.

Key: Use this data as a springboard—**celebrate** wins, **strategize** interventions, and **reinforce** conceptual depth.

The Educator's Role: Better Than Ever

Deepen Understanding & Application

AI excels at clarifying how to solve an equation; you help students connect it to real-world contexts, weave critical thinking, and reflect on outcomes. If the AI tutors them on grammar, you show them how great writing can transform ideas and change hearts.

- **Mini-Conferences**: Pull aside each student briefly after they work with AI. Ask how they felt about the day's practice, what they learned, and where they might apply it.

Strengthen Emotional Support & Motivation

Because AI handles the "drill-and-correct," you have **more time** to focus on rapport, emotional well-being, and creative engagement—exactly the parts of teaching that keep you and your students energized.

- **Encourage Growth Mindset**: Remind learners that AI feedback is just data—it's not a personal judgment. Mistakes are signals for next steps and growth opportunities, not evidence of failure.

Best Practices for Success

1. **Set Clear Usage Guidelines**
    - Students should understand that AI is a *tool*—not a place to cheat or mindlessly copy solutions.
    - Outline how to show their work or summarize their process, ensuring they're actively learning, not passively clicking.
2. **Blend AI with Group Work**
    - Follow up an AI practice session with pair or team activities. Let them **explain** what they learned, bridging solitary digital practice with peer interaction.
3. **Train Students to Question AI**
    - Just because AI says so doesn't mean it's always correct. Prompt them to find alternative methods or cross-check with references, fostering skepticism and thorough understanding.

4. **Integrate Real-World Projects**
    - If AI helps them master algebra steps, connect that to a physical building project, a robotics challenge, or entrepreneurial scenario. This cements knowledge beyond the screen.

Overcoming Common Concerns

- **"Doesn't This Encourage Laziness or Dependence?"**
    - Not if you require students to reflect on how they solved each problem. They can use AI for insights, but must articulate steps and reasoning, and lessons learned.
- **"What If They Just Click Through Without Thinking?"**
    - Build short, creative check-ins—like exit tickets where they detail new insights or "one thing AI taught me today."
- **"How Do We Keep Up with Tech Glitches or Updates?"**
    - Start with stable, well-known tools. Collaborate with other teachers or an IT point person for quick troubleshooting. Remind students to remain flexible when tech hiccups arise.

Key Takeaways

1. **AI Tutors Enhance Learning**: They handle routine tasks, offering 1:1 style support for each student.
2. **Teachers Remain Central**: You interpret data, address deeper

conceptual issues, and provide emotional and creative leadership.

3. **Practical Integration**: Use short AI sessions, track dashboards, then offer targeted mini-lessons or group activities.

4. **Holistic Approach**: Students must still demonstrate mastery beyond what the AI says—through reflection, projects, and real dialogue.

Next Up: In **Chapter 5**, we'll explore "Rethinking Assessment—Moving Beyond Standardized Testing" and discover how AI-driven adaptive quizzes and project-based evaluations can eliminate stress, highlight authentic progress, and open new dimensions for measuring understanding. Let's harness AI to **stop** grading kids as if they're identical, and **start** seeing who they truly are as individual learners and thinkers.

5

Beyond Bubble Sheets: Reinventing Assessment with Adaptive AI and Real-World Projects

—Stop rewarding memorization—start measuring creativity, real thinking, growth, and authentic mastery

For far too long, education has been trapped in a cycle of standardized tests—multiple-choice exams that prize memorization and uniformity. While this approach was once considered "fair," it often undermines the core purpose of teaching: **fostering genuine understanding, creative problem-solving, and the ability to apply knowledge in**

real-world contexts. AI can break us free from these outdated one-shot evaluations, replacing them with more nuanced, adaptive, and skill-based measures that paint a fuller picture of each learner's journey.

From One-Size-Fits-All Exams to Real-World Mastery

Traditional standardized tests rarely capture the depth of learning or the diverse strengths students bring to the table. A top scorer might simply be good at test strategy, while a quieter creative mind struggles to express insights in a time-limited format. The result? Stress, disconnected content cramming, and a teacher's scramble to "teach to the test."

1. **Limited Feedback**: Students often wait weeks for results, losing the chance for immediate correction or reflection.

2. **Unnatural Benchmarks**: Everyone is measured against the same rigid scale, regardless of pace or personal growth.

AI-driven assessments, however, adapt to each learner in real time. They track performance across multiple sessions, not just a singular exam day, and highlight conceptual understanding over brute memorization.

How AI Transforms Assessment

1. Adaptive Quizzes & Micro-Feedback

AI-Enhanced Tools

- **Century Tech** or **Cognii**: Assess each student's current level, then tailor questions in real time. If they flub a concept, the AI offers extra practice or simpler tasks.

- **Quizizz AI**: Creates dynamic, game-like quizzes that auto-scale difficulty and provide instant results.

Benefit: Instead of red marks on a test days or weeks later, learners see their mistakes *right away* and can address them immediately. Teachers get daily snapshots of class progress—no more guesswork on who's struggling.

2. Open-Ended AI Evaluations

Real Thinking Over Rote Memorization

A typical multiple-choice question rarely shows whether a student can write persuasively or solve an open-ended math problem. AI, however, can score or provide feedback on more complex student work—essays, short answers, or coding tasks—by analyzing structure, coherence, and even creativity markers.

- **Grammarly or Quillionz**: Offers line-by-line suggestions, prompts deeper revision, and fosters stronger writing skills.

- **AI in Math/Science**: Moves beyond "right vs. wrong" to check process, logic, and alternative solutions.

Benefit: Students practice *authentic expression*—from constructing arguments to designing solutions—while AI supplies immediate guidance that spurs iteration and growth.

3. Project-Based & Portfolio Assessments

Demonstrating Mastery Across Time

Rather than a single test, AI can track each student's evolution on complex, long-term projects—like building a simple video game, creating a scientific experiment, or writing a research paper. Teachers can see how learners adapt, refine, or collaborate across weeks or months.

Implementation

- **Regular Checkpoints**: AI logs progress, flags stumbling blocks, and notes innovative leaps.
- **Holistic Rubrics**: Educators evaluate final products and the journeys behind them—like conceptual depth, leadership in group tasks, or creative problem-solving.

Benefit: Students show who they are as *thinkers* and *innovators*, not just test-takers. Teachers assess real capabilities—critical thinking, resourcefulness, perseverance—beyond a bubble sheet's narrow lens.

Steps to Implement AI-Based Assessments

1. **Identify Subjects for Pilot**
    - Maybe start in one grade or in a skill-based class (like essay writing or pre-algebra). Gradual implementation ensures you can gather data and refine.
2. **Blend AI with Teacher Insights**
    - Let AI handle short quizzes, writing feedback, or conceptual checks. Meanwhile, you invest time in group projects, deeper discussions, and "conference style" feedback that addresses empathy, creativity, and broader skill sets.
3. **Teach Students to Read AI Feedback Critically**

- AI suggestions aren't infallible. Encourage them to question or rewrite any AI-based recommendation, building confidence and responsibility for their own learning.

4. **Go Beyond Grades**

   - Introduce badges, skill certifications, or digital portfolios. Colleges and employers are increasingly valuing not what you say, but what you can show. A portfolio, or cohesive record of real achievements and content you created, more than a letter grade or test percentile.

Common Concerns

- **"We Don't Have Time for Long Projects!"**
  - Freed from hours of repetitive grading, educators can reallocate that time to design integrated project experiences and reflect on AI-provided data.
- **"It's Hard to Measure Soft Skills Like Creativity or Leadership!"**
  - Combine teacher rubrics for intangible skills with AI data on student progress in collaborative or imaginative tasks. AI can design a rubric for you. Entrepreneurship is the ultimate future-proof skill.
- **"Parents Want Traditional Scores!"**
  - Show them how continuous AI-based feedback is more accu-

rate, less stressful, and fosters real growth. Provide side-by-side comparisons of old vs. new assessment results to illustrate deeper mastery. Show how AI is increasing the score, while fostering enthusiasm and a genuine love of learning, and see how this positively impacts their academic performance and behavior at home as they are increasingly empowered to learn and excel, at the specific level that's right for them. This gives kids confidence and sets kids up for success across all dimensions of life.

Key Takeaways

1. **Standardized Testing Alone Falls Short**: It measures too little, too late, often warping teaching priorities.

2. **AI Assessments Offer Real-Time Adaptability**: Students see immediate improvement paths, teachers see daily progress snapshots.

3. **Open-Ended & Project-Based**: AI can handle factual accuracy checks, leaving you to focus on nuanced creative or collaborative aspects.

4. **Integrate AI Ethically**: Always combine algorithmic scoring with human judgment, group tasks, and robust discussions to ensure fairness and depth.

5. **Drive Genuine Understanding**: Instead of cramming for an exam, learners engage in iterative problem-solving that hones the skill sets they truly need.

Next Up: Chapter 6 – Teaching Entrepreneurship: The Ultimate Future-Proof Skill. We'll explore how AI tools can spark your students' entrepreneurial spirit—teaching them to develop ideas, test assumptions, pitch solutions, and adapt to an ever-evolving world. Time to leave rote memorization behind and ignite the innovators in every classroom.

6

Entrepreneurship Unleashed: How AI Tools Help Students Launch and Lead

—The Ultimate Future-Proof Skill: Guide students to spot opportunities, research markets, solve real problems, & innovate

For years, career readiness meant preparing students for established jobs. But in an economy reshaped by AI and rapid disruption, **creating** new paths might be more valuable than merely filling existing roles. That's where **entrepreneurship education** comes in—teaching students not only to think like problem-solvers, but also to channel their curiosity, resilience, and big ideas into tangible ventures. And yes, AI can supercharge that journey.

Why Every Student Benefits from Entrepreneurial Thinking

1. A Dynamic Job Market Requires Adaptability

Jobs once considered "safe" can vanish overnight; new opportunities emerge where resourceful minds seize them. Entrepreneurs thrive on uncertainty, turning challenges into stepping-stones. By the end of this year, 90% of all code will be written by AI. Within the next 5 years, the World Economic Forum forum estimates 85 million jobs will be disrupted, but that 97 millions new AI-enhanced jobs will emerge for learners willing to upskill and adapt. Our kids are entering a world that is constantly evolving and being disrupted, and the most valuable skill we can impart is adaptability and entrepreneurship.

- **Classroom Example**: Students identify a local community issue—lack of healthy lunch options, say—and create a mini startup selling nutritious snacks. AI-driven marketing helps them figure out the best angles, but students plan budgets and connect with real users.

2. Financial Literacy & Business Skills Aren't "Extras"

In any AI-saturated future, navigating money, understanding supply and demand, and pitching an idea become critical life skills.

- **Positive Ripple**: Kids build confidence, collaboration, and negotiation skills—beneficial whether they found a company or lead a project within a larger organization.

How AI Fuels Student Entrepreneurship

AI-Driven Brainstorming & Validation

Free Tools like ChatGPT or Copy.ai can generate product name ideas, marketing taglines, or possible customer segments. Instead of floundering in blank-page syndrome, students get a quick spark—yet must refine it using human creativity and real data.

- **Activity**: Have students propose a business concept (e.g., an eco-friendly school supply kit). Let them query AI about potential target demographics or brand ideas. Then encourage them to pick or reject suggestions based on actual user surveys.

Market Research & Competitive Analysis

AI scrapes public data to highlight trends, competitor moves, or consumer demands. Even free versions of ChatGPT or Perplexity AI give kids a taste of genuine market intel, spurring them to iterate faster.

- **Outcome**: Students see how real entrepreneurs identify gaps, test assumptions, and adjust strategies—developing an entrepreneurial mindset that stands beyond classroom theory.

Prototype Building & AI Tools

No-code app builders or e-commerce sites with AI components allow quick MVPs (Minimum Viable Products).

- **Implementation**: Students design a simple landing page or chat-based service. AI helps with copywriting, color schemes, or layout. *But the student's core idea* remains front and center.

Integrating Entrepreneurship in Everyday Teaching

1. **Small Business Projects**
   - **Math Class**: Calculate costs, set pricing, track potential profits.
   - **Writing/English**: Craft compelling pitch decks or brand narratives with Canva or Gamma.
   - **Art/Design**: Develop logos, product visuals, or promotional graphics.
2. **AI-Based Shark Tank**
   - Students form mini teams, use AI for competitive research or marketing materials, then present to a panel (could be teachers, parents, or local professionals).
3. **Cross-Curricular Collaboration**
   - STEM students handle product functionality, Humanities-minded kids shape storytelling or marketing angles, and AI links them—pulling real data or optimizing text and positioning.
4. **Financial Literacy 101**
   - Combine AI-based budgeting apps with exercises on taxes,

revenue, and expenses. Let them see how minor decisions shape the viability of a startup idea.

Overcoming Resistance and Building Confidence

- **"But Our Students Are Too Young/Too Busy!"**
    - Start micro: a single class period each week for "idea labs." Even short bursts can stoke entrepreneurial sparks. Students can and will do amazing things when artificial limits are removed and they're empowered to ideate.
- **"We Don't Have Funding or Tech Tools!"**
    - Leverage free or freemium versions of AI platforms (e.g., local library resources, trial periods). Focus on concept demos rather than full-scale product releases. Perplexity.ai is a great resource for kids.
- **"Students Might Rely on AI for Every Aspect!"**
    - Encourage them to validate AI-driven suggestions via real user surveys or discussions. The mission: AI advises, humans decide. Trust but verify.

Key Takeaways

1. **Entrepreneurship is Future-Proof**: In this AI era, those who create solutions or spot and adapt to opportunities are the most valuable and future-proof.

2. **AI as a Catalyst, Not a Crutch**: Students can brainstorm, conduct quick market checks, or refine pitches faster, but still must do the interpretive and creative heavy lifting.

3. **Practical Activities**: Shark Tank challenges, mini e-commerce experiments, and cross-subject collaboration bring entrepreneurial ideas to life.

4. **Cultivating Leaders**: Entrepreneurship fosters grit, ownership, and the capacity to pivot in uncertain times.

Next Up: **Chapter 7 – AI & Coding: How Every Student Can Learn to Build with Technology**. We'll explore how even non-techy teachers can help kids embrace programming fundamentals, with AI removing syntax drudgery and unleashing unstoppable creativity. Let's turn your classroom from passive tech users into dynamic creators and innovators, empowered to bring their dreams to life.

7

Yes, They Can Code: AI Makes Tech Skills Fun and Accessible

— Empower students to transition from passive users to fearless creators, aided by AI's capacity to simplify and accelerate the coding journey.

Coding is often seen as a mystical or intimidating skill – full of cryptic syntax and complex error messages. Yet, in today's AI-driven landscape, **basic programming knowledge** opens doors for creativity, problem-solving, and entrepreneurial thinking across every career path. Thankfully, AI-based coding tools now reduce the friction of memorizing esoteric commands, allowing students to **focus on logic, design, and bringing their innovative ideas to life** instead of typing out every bracket or semicolon by hand.

Why AI Simplifies the Coding Roadblock

1. Instant Feedback Eases Frustration

Classic coding lessons can be painfully slow: type code, run it, sift through error logs. AI speeds this up by highlighting potential issues or automatically generating sample code, turning intimidation into curiosity.

- **Example**: In Replit's AI mode, a student types a request—"Build a Python program to track daily school lunch orders"—and sees starter code appear. They can tweak or inquire "why?" to deepen learning.

2. It Shifts Learning from Syntax to Problem-Solving

Instead of agonizing over punctuation and syntax, students reflect on **how** they want the program to behave, **why** certain logic matters, and **what** features to add next. AI handles basic structure, letting them concentrate on conceptual understanding and real-world application.

- **Outcome**: Freed from rote memorization, students gain the resilience and willingness to tackle advanced challenges (like building small websites, games, or interactive apps).

Classroom Approaches for AI-Assisted Coding

Approach 1: Small Daily Coding Prompts

Assign quick, "bite-sized" tasks using tools like **Scratch AI** or **Code.org** so students get repeated, low-stakes exposure to iterative coding.

1. **Prompt**: "Use AI blocks in Scratch to animate a cat dancing across the screen."

2. **AI-Generated Stubs**: Scratch might propose event triggers or loops.

3. **Student Customization**: They choose movements, backgrounds, or add fun twists.

Benefit: Everyone sees immediate results—reducing fear and stoking creativity.

Approach 2: Integrate Coding Across Subjects

1. **Math**: Program small geometry simulations or create a "fractal" visual using AI scripts that automatically handle the complex calculations.

2. **Science**: Model basic physics (like projectile motion) in a coded environment that AI helps debug.

3. **Language Arts**: Code a simple text-based adventure where students input story branches—an AI extension can suggest alternative endings or narrative arcs.

Key Tip: This underscores coding as a universal problem-solving language, not just a "tech elective." Tech skills are no longer required to bring your, or your kids' creative ideas to life.

Approach 3: Project-Based Collaboration

Encourage pairs or teams to conceive projects—a health-tracking app, an AI-driven flashcard tool, or a mini game—and let them rely on AI for code scaffolding.

- **Student Roles**: One might be the "designer," another the "coder," others the "researchers" or "documenters."
- **AI Input**: Tools like **ChatGPT, Claude,** or **GitHub Copilot** can auto-generate initial code or fix bugs.
- **Reflection**: Have the group discuss how AI contributed, where they had to override or refine, and how it affected their problem-solving.

Outcome: Students experience the real-world dynamic of "tech + creativity + team synergy," preparing them for future collaborative workplaces.

Overcoming Common Coding Hurdles

- **"I'm Not a Computer Science Teacher!"**
    - Modern AI platforms are user-friendly for educators with no coding background. You facilitate and let the tool handle deeper syntax. Start with simpler environments like Scratch or Code.org's AI labs.

- **“Won’t Students Just Depend on AI?”**
  - Proper guidelines ensure they explain how they integrated AI suggestions or tested modifications. Emphasize iterative design: “Here’s the AI’s starter code—how can we make it more interesting or fix its logic flow?”
- **“We Don’t Have Enough Devices?”**
  - Station rotations or shared-laptop pairings can still work. Some students do offline design, ideation or other tasks while others code. Everyone cycles through digital access.

Key Takeaways

1. **AI Lowers the Coding Barrier**: Syntax woes vanish, letting learners concentrate on logic, structure, and imaginative applications to bring their dreams to life.
2. **Cross-Subject Application**: Merge AI coding with math, science, language arts, or social studies to deepen understanding and spark creativity.
3. **Emphasis on Process Over Syntax**: The real skill is in *how* learners solve problems, *why* they choose specific logic, and *what* creative expansions they make. It's about asking good and clear questions, and developing clarity of vision rather than syntax drudgery. AI has leveled the playing field so anyone, irrespective of background or experience, can bring their creative ideas to life, faster and more effectively than ever.

4. **Teacher Support + AI Tools**: Even non-tech-savvy educators can lead big wins by allowing AI to handle the grunt work, while they supervise student progress, reflection, and real learning outcomes.

Next Up: Chapter 8 – Communicate to Captivate: Amplifying Public Speaking and Leadership via AI Support. It's time to see how AI helps with public speaking planning and feedback while you nurture authenticity, spontaneity, and leadership traits that technology alone can't replicate.

8

Communicate to Captivate: Amplifying Public Speaking and Leadership via AI Support

— Use AI prompts for structure—then add your human spark to engage hearts, inspire action, and cultivate tomorrow's leaders

As AI increasingly automates rote tasks—data entry, calculation, even initial content creation—**human communication** grows in importance. When your students can speak persuasively, share stories that move an audience, and navigate nuanced conversations, they hold an enduring competitive edge. AI can't replicate genuine empathy or spontaneity, but it can **assist** in the more mechanical aspects of writing a speech or analyzing one's presentation style to optimize it for maximum impact.

In this chapter, we'll explore how AI tools help your class sharpen communication skills. You'll see how to harness data-driven feedback for clarity, pacing, and structure—while retaining the **very human** warmth and presence that resonates in any speech or leadership moment.

Why Communication & Public Speaking Remain Vital in an AI-Driven World

1. **Leadership & Influence**: Robotically generated text can't replace the emotional resonance of a well-delivered human speech. Leaders who clearly articulate vision—teachers, students, CEOs—have a lasting impact.

2. **Collaboration & Team Building**: Effective communicators unify groups, handle conflict, and ensure that the best ideas come forward, even if AI provides raw data or suggestions.

3. **Adaptability**: While AI scripts can assist, real-time Q&A or impromptu debates demand human spontaneity and emotional intelligence.

Teacher Insight:

Students who practice public speaking early develop confidence, empathy, and higher-level thinking—regardless of their future career paths.

How AI Enables Public Speaking and Presentation Skills

1. AI-Assisted Speech Drafting

Tools like ChatGPT or Copy.ai can quickly generate outlines or entire scripts based on prompts. Instead of blank-page anxiety, students gain a solid starting point.

- **Class Example**: A 10th-grade humanities class is preparing for a debate on environmental policies. Students ask ChatGPT, "Act as an expert environmental policymaker. Outline a 3-minute opening statement defending renewable energy subsidies," then refine the content with personal flair and local data. I also like to ask ChatGPT to leverage principles of Smart Brevity (from Axios) to deliver key points as crisply and compellingly as possible.

Key Benefit: They save time on structural grunt work, focusing on purpose, logic, and rhetorical style.

2. Real-Time Feedback on Delivery

Apps such as Yoodli or Speeko AI analyze vocal tone, pacing, filler words, and clarity, offering specific pointers.

- **Implementation**: Students record themselves presenting a short speech. The app flags "ums," monotone sections, or rushed conclusions.

- **Class Follow-Up**: Pair up students to cross-check AI's remarks with peer impressions—what did they notice about body language or emotional connection?

Key Benefit: AI's data-driven insights address technical pitfalls fast, while human listeners address intangible qualities like passion or empathy.

3. Enriching Group Discussions and Debates

Structured Roles: For collaborative projects, assign an "AI Summarizer" role—someone who uses an AI tool to compile group notes or refine a final statement. This fosters accountability for fact-checking, ensuring the group's ideas come through powerfully. You could host a Zoom meeting from the classroom and have Otter.ai or Read.ai summarize everyone's comments and provide engagement and performance analytics as well. The possibilities are endless.

Note: Students remain in charge of verifying correctness and deciding how best to phrase ideas for persuasive or informational impact.

Balancing AI with Authentic Human Expression

Don't Let AI Flatten Your Voice

Though AI can produce neat, functional text, the emotional texture—that personal story, humor, or vulnerability—belongs to the speaker. Educators and kids can actually teach AI their "voice" or "style" and have it increasingly "learn" their preferred communication quirks. This is a more advanced skill, but it streamlines the creation process even further.

- **Classroom Tip**: Ask each student to add a personal anecdote or a surprising angle the AI never would've found in its data bank. They must flesh out the emotional "why" behind their topic.

Question "Just-So" Answers

AI often defaults to safe, standard arguments. The "general consensus," unless you specifically instruct it otherwise, for example, to act as a specific personality. If students only rely on the general consensus, they might miss deeper perspectives. You can train kids to prompt AI to "act as an expert in Y, and present 10 arguments for and against topic Z" to get a broader range of perspectives.

- **What to Do**: Encourage them to challenge AI's statements, expand or invert them, and gather real-life examples. The result is a more original, authentic narrative.

Activities & Implementation Ideas

1. **AI Draft vs. Human Rewrite**
    - Students generate an AI speech draft, then craft a revised version focusing on voice, emotional hooks, and rhetorical devices.
    - **Discuss**: Which parts felt robotic? Which edits made it more relatable?
2. **Filler Word Challenge**
    - Have students deliver a quick talk—maybe 1–2 minutes on a hobby—and record it in Yoodli. The app flags filler words such

as "like," "um," or "so."

- **Reflection**: They share their experiences with the app's feedback, then set a personal reduction goal for next time.

3. **Debate with AI-Enhanced Prep**

   - Split the class into teams, each aided by AI for research. The system suggests arguments and counter-arguments.

   - They refine these points with real data, personal stories, or quotes. Then they conduct a live debate, practicing spontaneous rebuttals and checking for emotional resonance.

Overcoming "Fake Voice" Concerns

- **"Won't AI ruin originality?"**

  - Remind your class that AI outputs are a *rough scaffold*. True persuasion demands sincerity, personal anecdotes, and emotional nuance.

  - They can also incorporate local references or personal experiences the AI can't glean from a general data set.

- **"Our kids need organic skill-building, not AI crutches."**

  - Absolutely, so consider AI as a *mentor* that spots mechanical or structural weaknesses quickly. Students still must internalize lessons by repeating and practicing in real scenarios.

Key Takeaways

1. **Speech & Leadership**: Communication remains a top skill; AI can shape outlines or track vocal mannerisms, but genuine expression is yours alone.
2. **Technical & Emotional Feedback**: Tools like Yoodli or Speeko highlight pacing, filler words, or monotone segments. You or peers address authenticity and emotional punch.
3. **Authenticity Over Generic**: Don't let AI drafts overshadow personal voice. Students must bring unique stories and angles to truly captivate audiences.
4. **Classroom Integration**: Quick practice sessions, group debates, or final project presentations can incorporate AI-based evaluation. The result? Confident communicators ready for stages of any size.

Next Up: Chapter 9 – A Blueprint for Burnout-Free Schools: How AI Reduces Repetition and Lets Teachers Thrive. Let's move from individual classroom tactics to a broader vision of how entire schools can unify around AI systems—relieving staff stress, uplifting morale, and creating equitable, engaging learning experiences for all.

9

A Blueprint for Burnout-Free Schools: How AI Reduces Repetition and Lets Teachers Thrive

—Design campus-wide AI integration that boosts morale, fosters creative energy, and provides equitable opportunities for all learners.

Individual teachers experimenting with AI can trigger meaningful transformations in their classrooms. Yet the greatest impact emerges when **entire schools or districts** adopt a thoughtful AI strategy—backed by leadership support, professional training, and a shared vision. In such an environment, **repetition and drudgery decline**, replaced by **collaborative innovation** and a powerful sense of purpose. Welcome to a new educational blueprint where staff feel energized instead of overworked, and students receive an equitable springboard toward success.

Why School-Level AI Initiatives Matter

1. Consistency & Scalability

When each teacher uses a different AI platform—or none at all—students experience inconsistent experiences. By aligning tools, training, and policies, the entire faculty can streamline tasks and unify best practices.

- **Example**: A middle school adopts Khan Academy AI across all math classes. Teachers compare dashboards, share techniques, and help each other interpret data for targeted interventions. Students move from class to class with the same adaptive tool, reducing confusion and tech friction.

2. Culture of Innovation

A single pilot might dazzle a few classes, but a culture shift means teacher collaboration, leadership buy-in, and parent/community understanding. AI stops being a novelty and becomes an integral part of *how* learning happens daily.

- **Outcome**: Staff no longer see AI as an optional fad but as a resource to cut workload and boost outcomes—freeing them for problem-solving, enrichment, coaching, and mentoring students.

The 5 Pillars of an AI-Powered School

Pillar 1: AI-Powered Personalized Learning

- **Adaptive Platforms**: Deploy consistent, proven AI tutors (like Khan Academy AI, DreamBox, or Century Tech) across relevant subjects.
- **Shared Metrics**: Teachers and administrators monitor uniform dashboards, spotting trends and reacting quickly to improvement or struggle.

Benefit: No child falls through the cracks because everyone's performance is tracked individually in real time.

Pillar 2: AI as Your Teachers' Assistant

- **Grade Automation**: Tools like Grammarly or Quillionz accelerate writing feedback, while certain math AI systems handle step-checking.
- **Lesson Plan Suggestions**: ChatGPT or specialized solutions can churn out structured outlines or quiz banks in seconds.

Outcome: Teachers spend less time on routine tasks, more on creative lesson design, advanced concept teaching, or one-on-one student connections.

Pillar 3: Skill Development—Entrepreneurship, Coding, & Leadership

- **Dedicated Modules**: Launch mini bootcamps or cross-subject projects where students harness AI to create and innovate.
- **Labs/Clubs**: Offer after-school "AI & Coding" clubs, "Shark Tank" competitions, or hackathons that merge entrepreneurial thinking with technology.

Impact: Students see school as a launchpad for real-world skills—coding prototypes, building businesses, refining leadership traits.

Pillar 4: Project-Based STEM & Hands-On Learning

- **AI-Integrated Labs**: Virtual science explorations (Labster, PhET) remove equipment limitations and expand experimentation.
- **Maker Spaces**: Pair physical tools (3D printers, robotics kits) with AI design suggestions, letting students iterate quickly.

Result: Teachers coordinate interdisciplinary tasks—English classes do storytelling for an AI-coded game, math classes handle calculations, science classes test real data. Everyone's engaged in ideation, and they can see how they are empowered to bring their dreams to life.

Pillar 5: Student Well-Being & Emotional Intelligence

- **AI-Based Counseling**: Chatbots like Wysa or Woebot can offer basic mental health check-ins, though human counselors remain crucial.

- **Empathy Projects**: Use AI analytics to gauge social dynamics, so staff can intervene promptly if conflicts or isolation patterns emerge.

Objective: Nurture the "whole child" by merging emotional literacy with academic pursuits. Freed from busywork, teachers can be more present and supportive.

Steps to Implement AI at the School Level

Step 1: Teacher Training & Pilot Selection

- **Professional Development**: Offer hands-on workshops where teachers try AI for grading or generating practice sets.
- **Choose a Subject or Grade**: Pilot in, say, 6th-grade math for a semester—collect data, gather teacher feedback, refine usage.

Key: Involve teacher leaders from day one; teacher buy-in is everything.

Step 2: Ethical Policies & Data Privacy

- **Legal Compliance**: Adhere to FERPA/COPPA in the U.S., or relevant data protection laws elsewhere.
- **Transparency**: Let parents, students, and staff know how data is collected, stored, and used.

- **AI Literacy**: Teach kids how AI arrives at answers, encouraging them to question biases or limitations, ask better questions, and provide clear guardrails.

Essential: This fosters trust and prevents misunderstandings around "surveillance" or misuse of personal data, in addition to teaching kids the importance of asking good, targeted questions.

Step 3: Foster Collaboration & Peer Coaching

- **AI Mentors**: Identify at least one "tech-savvy" champion per department or grade to support colleagues.
- **Teacher-Led R&D**: Encourage short weekly meets to share AI tips or classroom wins. Document best practices in a shared hub.

Effect: Educators gain confidence, quickly troubleshoot issues, and celebrate successes, fueling a sense of unity and innovation.

Step 4: Monitor, Evaluate, Refine

- **Data Dashboards**: Principals or AI committees track metrics—student progress, teacher feedback, stress levels.
- **Feedback Cycles**: Survey teachers on time saved, new challenges, or professional satisfaction. Survey students on engagement and learning.

Result: Continuous improvement rather than a one-off experiment. Adjust the approach based on real evidence and teacher/student voices.

Real-World Snapshot

Case Study: A suburban middle school deploys AI-based tutoring (DreamBox for math, Grammarly for writing) across 6th–8th grades. They also adopt an AI app for quick reading comprehension checks. Teachers free up an average of 10 hours weekly from manual tasks. Stress dips, student confidence rises. Over a semester, test scores in math and writing show gains, but there's also an impressive **drop in teacher burnout** and an *upsurge* of creative projects teachers now have time and energy to implement.

Key Takeaways

1. **School-Wide AI Integration** Produces Greater Consistency & Unity: Students benefit from transformational tools, staff share strategies, and data is cohesive.

2. **AI Offloads Repetition** So Teachers Can Innovate: Freed from drudgery, educators can dive into advanced lessons, emotional support, or broader skill-building.

3. **Ethical & Transparent Rollout** Builds Trust: Data compliance, AI literacy for staff/students, and pilot-based expansions reduce fear and confusion and set educators and students alike up for success in the future.

4. **Collaborative Culture** Grows: AI mentors, shared best practices, and open reflection help everyone adopt progressive methods without feeling alone or overwhelmed.

Next Up: **Chapter 10 – AI at Home: Partnering with Parents to Supercharge Learning and Relieve Stress**. We'll explore bridging classroom and living room via AI-driven tutoring, reading, and curiosity—while staying mindful of screen time, privacy, and real-world balance.

10

AI at Home: Partnering with Parents to Supercharge Learning and Relieve Stress

—Bridge classroom and living room via AI-driven tutoring, reading, and curiosity—while preserving healthy tech boundaries.

School can accomplish a great deal, but the journey doesn't end when students leave campus. Parents play a massive role in reinforcing, extending, and personalizing what kids learn—and AI can be a potent ally. Rather than piling extra pressure on families, **well-chosen AI tools** help parents monitor progress, offer targeted support, and create a more engaging environment without feeling like an additional burden. This chapter shows educators how to **collaborate with parents** so home-based AI learning complements class instruction and fosters genuine growth.

Why Home AI Integration Matters

1. **Extended Practice & Reinforcement**
   Students who get adaptive practice or instant feedback at home stay more confident and connected. They don't arrive at school the next day feeling lost or overwhelmed.

2. **Parent Empowerment**
   AI dashboards can give parents real-time snapshots of their child's strengths or gaps. Instead of vague "How was school?" queries, families can dive into specific challenges or celebrate improvements.

3. **Reduced Homework Battles**
   AI-based tasks often feel more interactive—turning routine homework into a gamified experience or quick skill check. Less nightly friction means parents shift from "homework police" to supportive guides.

Teacher Insight:

When families embrace AI responsibly, stress on both sides diminishes, letting educators and parents share a consistent message: "We're on the same team, ensuring you learn in multiple, fun ways."

AI Tools Families Can Use at Home

1. Personalized Tutors & Apps

- **Khan Academy AI**: Offers self-paced modules, combining videos with adaptive quizzes.
- **DreamBox**: Engaging math games that adapt in real time to a child's performance.
- **Duolingo**: Reinforces foreign language skills with quick daily challenges, often spiced up by AI-based question variations.

Classroom Tie-In: Provide parents with recommended sessions or highlight the precise concept a student struggles with, so the app automatically tailors practice.

2. Reading & Writing Support

- **Google Read Along**: AI listens to kids read aloud and corrects pronunciation gently.
- **Grammarly**: Instantly flags grammar and clarity issues for older children's essays or creative pieces.
- **Quillionz**: Generates comprehension questions from any text, turning reading time into deeper engagement.

Outcome: Instead of parental guesswork—"Did you read well? Did you understand the chapter?"—AI supplies real-time evidence of fluency or textual grasp.

3. Interactive Projects & Curiosity Builders

- **Scratch & Scratch Jr.**: Introduce coding logic in playful ways; parents can watch or join the creativity.
- **StoryLab.ai**: Helps young storytellers brainstorm plot lines, characters, or endings, bridging writing with imagination.
- **Little Inventors**: Some tools use AI prompts to challenge kids to design mini inventions at home, encouraging hands-on building.

Benefit: Merges structured learning with free-form creativity, stopping short of turning the home into a rigid extension of school.

Best Practices for Teachers to Facilitate Parent Engagement

1. **Offer Clear Guidance**
    - Provide a brief list of recommended AI tools, each with a short how-to. Outline potential daily or weekly usage. Emphasize moderate screen time and purposeful use.
2. **Sync Class Goals with Home Tasks**
    - If your 5th graders are tackling decimals, suggest a 15-minute "DreamBox decimals module" each evening. The child arrives at school with fresh practice and the AI's instant feedback, letting you jump into deeper instruction.
3. **Encourage Reflection**

- Parents can ask kids, "What did you learn from the AI app today? Any surprises or tricky spots?" This fosters ownership rather than passive usage. And they can view their kid's dashboard so they can ask inform questions and explicitly see progress and understanding, in real time.

4. **Foster Digital Literacy & Privacy**

   - Remind families to keep logins secure, read data policies, and teach kids to cross-check AI answers. If the platform offers parental dashboards, show them how to interpret results.

Balancing AI with Real-World Activities

While AI apps can be transformative, an over-reliance can erode other crucial experiences—like outdoor play, arts, or plain downtime. The magic lies in synergy:

- **30–45 Minutes Max**: For younger kids, short AI sessions suffice. Let them remain mentally fresh without turning "digital practice" into an electronic babysitter.
- **Offline Extensions**: If the AI tool taught basic fractions, have the child measure ingredients while cooking dinner. If it introduced an interesting historical figure, discuss how that person's legacy appears in your local community.

Teacher Tip:

Periodically check in with parents about how kids are blending AI lessons with everyday life—are they exploring new hobbies sparked by a coding challenge, or are they stuck on screens?

Mitigating Common Pitfalls

- **"Parents Might Over-Rely on AI for Homework Help."**
    - Encourage them to see the AI as an assistant, not a replacement. Students should still articulate how they arrived at solutions, not simply regurgitate AI text.
- **"Some Families Lack Devices or Strong Internet."**
    - Suggest offline or low-tech alternatives—e.g., library visits, borrowed district devices, or simplified phone-based versions if laptops aren't available.
- **"Screen-Time Concerns."**
    - Provide quick guidelines (e.g., a set window each day, or tool usage must follow real-world chores or creative pursuits) and emphasize the role of adult oversight and conversation about what's learned.

Key Takeaways

1. **AI at Home Extends Learning**: Students receive ongoing practice, immediate feedback, and a sense of continuity between

school and family support.

2. **Parents Feel Involved, Not Overloaded**: Real-time data plus user-friendly apps reduce guesswork about a child's progress.

3. **Healthy Integration**: Emphasize short, purposeful sessions balanced by offline experiences—ensuring kids develop a well-rounded skill set and emotional grounding.

4. **Teacher's Role**: Guide families on recommended apps, usage boundaries, and synergy with class goals. This partnership fosters stronger connections and more motivated learners, and ensures this isn't "just another things kids, parents and teachers have to do" but a time and energy saver and motivator for all parties.

Next Up: **Chapter 11 – The Future of Education: What's Next?** With the disruption in the Department of Education, things are more uncertain than ever. Here, we explore opportunities to uncover money for key initiatives without introducing risk, so you can stay in control and empowered to pursue key projects despite this current environment of uncertainty and disruption.

11

The Future of Education—How To Fund Your Dream Initiatives

—With the disruption of the Department of Education, how to find funding & prepare your school for the next decade

The foundation of American education is shifting—and not slowly. The recent disruption across the Department of Education (DOE), from funding stalls to staffing instability, has created deep uncertainty. But in that uncertainty lies a rare window of opportunity for those willing to seize it.

For decades, we've tried to patch the cracks in a century-old school model—one designed for an industrial economy, not an AI-driven one. Now, we have the opportunity to rethink the system from the ground up.

This chapter isn't about doom and gloom or politics. It's about taking a fact-based look at ramifications, and how we can work within this disruption to create something better. It's about what's next—and what's *possible*.

Big Picture:

Fully shutting down the DOE would significantly shift power to states, reducing federal oversight and offering flexibility and efficiency at the local level, but it also carries significant risks, including increased inequality, weakened national standards, and compromised college accessibility.

Ramifications

- **Funding shifts:**
  $80+ billion federal education funds redistributed or eliminated, impacting state budgets.

- **Decentralization:**
  States gain full authority over education policy, curriculum, and standards.

- **Loss of uniformity:**
  Educational standards and outcomes could become uneven across states.

- **Higher education disruption:**

Pell Grants and federal student loans would need new administrators or face elimination, limiting college access.

- **Civil rights protections:**
  Reduced federal oversight could weaken anti-discrimination enforcement.

Benefits

- **Local flexibility:**
  States tailor education policies directly to local needs.
- **Reduced bureaucracy:**
  Potential taxpayer savings by cutting federal administrative overhead.
- **Enhanced parental choice:**
  States may expand vouchers, charter schools, and alternatives to public education.

Risks

- **Educational Inequality:** Risk of widening the gap between wealthy and lower-income communities.
- **Inconsistent Quality:** Loss of uniform academic standards could weaken U.S. competitiveness globally.
- **Reduced College Access:** Potential decline in affordable higher education without federal student aid.

- **Weakened Civil Rights:** Less effective enforcement of laws protecting vulnerable student populations.

Bottom Line

Success hinges on states' capability to effectively manage education without federal support. Careful transition planning would be essential to avoid exacerbating inequalities and harming national educational outcomes.

The Challenge: A System in Flux

When a central agency like the DOE falters, schools, districts, and educators are left exposed. Federal grants stall. Oversight lapses. Policy guidance grows unclear. Educators on the ground face rising pressure—with fewer resources and less clarity.

But the deeper issue isn't just budget instability. It's that the traditional model of K–12 education no longer fits the world our kids are growing into.

Core issues:

- **One-size-fits-all pacing**: Forces all students through content at the same rate—ignoring both gifted learners and those needing help.
- **Seat time over mastery**: Rewards attendance and completion, not real understanding.

- **Outdated assessments**: Standardized tests measure memory, not the critical skills of today like creativity, problem-solving, or adaptability.
- **Burnout culture**: Teachers are overwhelmed in 30:1 classrooms. Parents are frustrated. Kids are disengaged.

While these are uncertain and scary times, it may also be an opportunity. This is the moment to move from disruption to design. To stop "tinkering around the edges" and finally give students and educators a model that reflects the world we live in—and the future we're entering.

The Opportunity: Rethinking School for a New Era

This is where bold leadership comes in. We don't need to wait for the system to fix itself. Many of the solutions already exist—and they work. It's just about knowing about them and finding funding for them

Let's break them down.

1. AI Tools for Personalization at Scale

AI is not about replacing teachers. It's about giving them *superpowers*—freeing up time, tailoring instruction, and supporting diverse learners all at once.

Well-implemented AI tools can:

- **Diagnose learning gaps in real-time**
- **Adapt instruction to each student's pace**

- **Free teachers from grading, prep, and tedious tasks**
- **Support multilingual learners and special education needs**

Platforms like **Khanmigo, Century Tech, DreamBox, JasperAI, and GrammarlyGO** are already proving that AI can reduce burnout while improving student outcomes.

When AI handles the repetitive tasks, teachers reclaim the time to build relationships, lead creative projects, and develop leadership and emotional intelligence—the things AI *can't* do.

2. The 2-Hour Learning (or similar) Models: Mastery Over Motion

Inspired by cognitive science and tested in real classrooms, the **2-Hour Learning** model shows that students can achieve *more* in *less* time—when learning is personalized, mastery-based, and AI-enabled. Students complete entire grades in 3 months, and score in the top 1% of their peers, consistently, in standardized tests. This is not a fluke. It's personalized AI coaching in action.

Key elements:

- 60–90 minutes of focused, adaptive learning via AI or curated content
- 30–60 minutes of real-world application (projects, discussion, reflection, entrepreneurship, life skills)
- Emphasis on *competency*, not seat time. Students can't move forward unless they've show at least 90% mastery. And they're finish-

ing grades in 3 months with just 2 hours of lesson time per day. It's staggering what can be achieved when artificial limits are removed from kids.

With this model, students move at their own pace—accelerating through areas of strength and getting targeted support in areas of need.

It's especially powerful for schools trying to protect instructional time while integrating arts, STEM, entrepreneurship, and life skills. It's liberating, and the results are incredibly compelling, but it's costly. How do districts who are already facing budget cuts and uncertainty afford these types of AI-powered teacher and student enablement solutions?

3. Cost-Saving Innovation: Funding the Future Without Waiting

The #1 pushback from districts and administrators? **"We love it, but we can't afford it."**

The secret is, the money to fund these initiatives is already hidden in your budget, you just don't know it yet.

Contingent cost-saving consultancies offer a radically different way to fund innovation:

- No upfront cost
- No risk to the school
- They only get paid from *realized savings*

These firms analyze your district's operations and *indirect costs* (from everything from transportation and facilities to tech procurement and utilities and other vendor-driven solutions), uncover inefficiencies, run RFPs without taking school or district administrator time, and renegotiate contracts to ensure monetary savings without loss in product or service quality, and redirect savings toward strategic goals like:

- AI integration
- Microcredential programs
- Entrepreneurship and life skills labs
- Teacher and student support systems and much-needed resources

These contingent cost-savings consultancies only get paid if they find savings and you decide to move forward, and they can uncover millions. The group I partner with has helped thousands of organizations uncover over $3B in hidden savings. If this is something you'd like to explore, and wish to pursue a free consultation, please visit LeadWithGold.com/save.

What could you do if you could free up 20% of your expenses for initiatives of *your* choosing to empower *your* teachers and students, without layoffs or reductions in product or service quality? This approach flips the narrative from **"we can't afford it"** to **"we can't afford *not* to."**

The Risks: What Happens If We Don't Adapt

Especially with the disruption of the DOE, if we continue clinging to the status quo, the cracks will grow:

- Growing disengagement and chronic absenteeism

- Declining reading and math scores
- Teacher shortages and morale collapse
- An even wider gap between schools that innovate and those that don't

And worst of all: another generation of students who feel unseen, unsupported, and unprepared for the world that awaits them.

The Reward: A System That Actually Works—for Everyone

Here's what happens when we get this right:

- Students finish school **with real skills**, not just transcripts
- Teachers feel **supported and empowered, not stretched thin and burned out**
- Classrooms become spaces for **creativity, not compliance**
- Equity gaps shrink, because **AI-driven personalized learning scales**
- Communities rebuild trust in schools as hubs of possibility, preparing their children to thrive in the world of tomorrow.

Next Steps: What You Can Do Today

Whether you're a teacher, school leader, parent, or policymaker—this is your moment to lead. Here's how to start:

1. **Pilot AI strategically**
    - Start with one subject, grade level, or group. Measure impact. Scale thoughtfully.
2. **Adopt a 2-Hour Learning (or similar) prototype**
    - Try it for one grade band. Monitor academic and engagement results. Share success stories.
3. **Explore cost-savings partnerships**
    - Engage a no-risk consultancy to fund your innovation roadmap from within existing budgets.
4. **Host a community conversation**
    - Help families understand how AI + personalization = *more humanity, not less.*
5. **Shift the culture from fear to possibility**
    - Remind your team: this isn't about adding more, its about saving time and energy and **doing differently—with purpose, clarity, and courage, so they have more time, autonomy, and energy for the things that matter most.**

Final Thought

The future of education isn't waiting for Washington. It's being built in your district, your classroom, your living room—right now.

We're no longer in the business of filling seats and checking boxes. We're in the business of igniting potential. And when we stop chasing the past, we can start designing the future our kids deserve.

Let this moment of disruption become the moment we chose to lead.

Next Up: Chapter 12 – Leading the AI-Driven Education Revolution. We'll wrap up by focusing on *you*, the educator or leader, and the steps you can take right now—no matter your starting point—to spearhead innovative, burnout-free classrooms and a future-ready learning community.

12

Lead the AI Revolution: A Teacher's Roadmap to Lasting Impact Without Burnout

– Empower your classroom, inspire your peers, and shape tomorrow's schools with confidence, vision, and renewed passion.

We've journeyed through a vision of schools where AI tools, personalized lessons, and new approaches to teaching become a powerful force—reducing teacher workload, igniting student engagement, and preparing everyone for a rapidly evolving world. But the crux of this transformation rests on **leadership**—not just from administrators, but from you, the educators who dare to try new ideas, collaborate with peers, and champion responsible AI usage.

This final chapter focuses on the **tangible steps** you can take—at the classroom, campus, or district level—to **become the catalyst** for an AI-powered, burnout-free educational environment.

1. Embrace a Pilot Mindset

Why Start Small?

Big, sweeping changes often trigger fear or pushback. By beginning with one class or one grade, you gather success stories and iron out kinks, setting the stage for broader buy-in.

Action Steps

- Identify a single unit—say a 4-week math module or an English writing project—to integrate an AI platform (Khan Academy, Grammarly, etc.).
- Collect baseline data: teacher workload hours, student quiz scores, and engagement levels. Post-pilot, compare results to illustrate tangible improvements.

Outcome:

You build a positive case study that encourages more teachers to adopt AI and more administrators to back your proposals.

2. Cultivate Teacher Collaboration & Peer Support

The Power of a Shared Vision

It's easy to feel isolated if you're the only one venturing into new tools. **Community** is the antidote. Form AI roundtables, schedule quick "tech share" segments in staff meetings, or create a digital hub where educators exchange tips and experiences.

Action Steps

- **AI Buddy System**: Pair up teachers—one with more tech comfort, one with less—to help each other.
- **Monthly Mini-Demos**: Each month, one teacher showcases a short success (like using AI for immediate quiz feedback). This fosters collective growth

Outcome:

The energy of shared discoveries multiplies, fueling a sense of unity and consistent progress across classrooms.

3. Engage Parents & Communities Early

Building Trust & Understanding

Parents who only hear "AI is in the classroom" might worry about screen time, data privacy, teacher displacement, or encouraging student laziness. Communication dissolves these fears by clarifying **how** AI is used (supporting, not replacing teachers), and how it benefits their children academically and emotionally.

Action Steps

- **Parent Workshops/Webinars**: Demonstrate a live example of an AI tutoring tool, show its feedback cycle, and highlight how it enables and empowers, students, parents, and teachers alike.
- **User-Friendly Guides**: Provide short "how-to" sheets or videos for families interested in using or monitoring the AI platform at home.

Outcome:

When parents feel informed, they become enthusiastic allies—reinforcing AI-based tasks, celebrating progress, and trusting the process.

4. Advocate for Ethical & Transparent Policies

Why Ethics Matter

AI in education involves data tracking, algorithmic decisions, and potential biases. Upholding **trust** requires strong ethical guidelines, including compliance with FERPA/COPPA in the U.S. or equivalent laws elsewhere.

Action Steps

- **Form an Ethics Committee or Task Force**: Teachers, admins, parents, and possibly tech experts discuss data governance, best practices, and pilot expansions, along with key guardrails for success.

- **Teach AI Literacy to Students**: Show them how AI outputs are generated, the risk of bias and how to mitigate it, and the importance of verifying facts.

Outcome:

A clear set of do's and don'ts fosters a safe and supportive environment, preserving faith in AI's educational potential.

5. Integrate AI Tools Into Official Curriculum & Professional Development

Making It "Normal," Not an Add-On

When AI remains a side project, it can easily fade. By weaving it into **district-approved curricula** and professional development (PD) requirements, you create an institutional framework that solidifies usage and invests resources in ongoing improvement.

Action Steps

- **AI & Entrepreneurship Modules**: Expand your existing business or STEM courses to include AI-based tasks—like building a real or hypothetical startup.
- **Professional Learning**: Encourage or mandate short PD activities on adaptive tutors, AI-based grading, or micro-credentialing.

Reward teacher-led discoveries with credit or recognition.

Outcome:

AI-based approaches become a natural, recognized part of teaching, with the support of the entire school ecosystem.

6. Document & Broadcast Wins

Motivate Others Through Concrete Results

Once a pilot or new approach shows positive outcomes (higher engagement, less teacher burnout, improved mastery), share it widely. Visual data, quotes from students, or personal teacher testimonies have persuasive power.

Action Steps

- **Write Case Studies**: Summarize how an AI writing tool halved your grading time, freeing you for creative lessons.
- **Present at Faculty, District, or Even Regional Conferences**: Champion your students' achievements and your own improved well-being.

Outcome:

Momentum builds—colleagues see your successes, parents feel proud, and administrators find justification for resource allocation.

7. Secure Funding for Your AI Vision

Why Funding Matters

Even modest AI rollouts—be it a new set of adaptive licenses or upgrading network infrastructure—can stretch tight budgets. With strategic planning, you can gather resources that bring your innovative plans to life without burdening teachers or administrators.

Action Steps & Strategies

1. **Start with a Clear, Measurable Pilot Plan**
    - Outline exactly how your AI pilot will improve reading scores or reduce teacher grading hours. Concrete goals appeal to potential funders—whether internal or external.
2. **Tap into Grant Opportunities & Foundations**
    - Look for STEM, digital literacy, or local education grants; align your proposal with their goals (equity, improved outcomes, teacher retention, etc.).
    - Build relationships by emailing or calling foundation contacts, personalizing your application rather than spamming generic requests.
3. **Revisit District Budgets & Funds, & Explore No-Risk Cost Saving Consultations**
    - Work with administrators to see if they are willing to en-

gage no-risk cost saving consultancies to cover AI-based PD or adaptive programs.

- Show that teacher time saved translates into better instruction and cost savings. For a free consultation, visit LeadWithGold .com/save.

4. **Engage Parents & Community for Crowdfunding**
   - Platforms like DonorsChoose let micro-donations fund specific classroom needs (headsets, app licenses, etc.).
   - Sharing success stories or real student achievements sparks emotional buy-in.

5. **Seek Tech Industry Partnerships**
   - Some companies offer pilot programs or sponsor AI-driven learning experiences. Approach them with a mutually beneficial pitch, highlighting brand goodwill and workforce pipeline benefits.

Navigating Funding with Integrity & Impact

- **Align Every Dollar with Educational Goals**: Ensure tools or devices directly address real pain points—like teacher burnout or student skill gaps.
- **Maintain Transparency & Accountability**: Publicize budgets, usage updates, and data to keep stakeholders' trust.
- **Avoid Over-Reliance on One Source**: Diversify grants, district

reallocation, and local sponsorships so one funding stream isn't make-or-break.

Outcome:

With a sustainable funding model, you anchor your AI initiatives in real impact rather than fleeting novelty—securely uplifting your entire school community.

8. Maintain a Human-Centered Focus

Beyond Data & Efficiency

Yes, AI can show a data dashboard or speed up tasks, but your biggest impact rests in building relationships, empathy, and moral guidance for your learners.

Action Steps

- **Daily Personal Connection**: Even if AI tracks progress, make room for short face-to-face check-ins with students each week. Show them they aren't just data points.
- **Reflective Practices**: Encourage journaling or class discussions about their AI experiences—victories, frustrations, and quirks. This fosters emotional intelligence and genuine teacher-student bonds.

Outcome:

Students see AI as a tool, not an end. Your leadership ensures personal growth remains at the core.

Putting It All Together

In the chapters leading here, you've explored:

1. **Personalized Learning** through AI, easing teacher burnout.
2. **Myth-Busting** clarifying how AI complements, not replaces, human instruction.
3. **Future-Ready Skills** (creativity, coding, entrepreneurship) that AI elevates.
4. **Assessment Reimagined** with adaptive quizzes and project-based evaluations.
5. **School-Wide Blueprint** unifying staff around AI integration.
6. **Parent Partnerships** bridging classroom and home-based AI synergy.
7. **Funding** strategies to sustainably support your vision.
8. **Human-Centered Focus** that keeps empathy and moral growth paramount.

Stepping into your leadership role means weaving all these strands into a coherent strategy, championing teacher autonomy and creativity, and forging an educational system that's **unafraid to innovate**.

Key Takeaway: By blending your passion and expertise with AI's efficiency and personalization, you create a **future-proof learning environment**—one where teachers are *energized*, students are *deeply engaged*, and the entire community sees education as a living, evolving adventure. Embrace the challenge, forge the path, and be the beacon of possibility that transforms your school into a model of AI-powered, humanity-centered teaching.

CONCLUSION

–*Carrying the Torch of AI-Driven Education*

As you close this book, realize you're standing on the threshold of **redefining education** for an era marked by exponential change. You've seen how AI can tailor lesson paths, reduce mechanical workloads, and free educators to focus on deeper human connections. You've learned how teachers, parents, and administrators can align to bring each student a richer, more personalized experience. You've discovered ways to find funding, navigate ethics, and scale up from small pilots to campus-wide transformations.

But real success hinges on **your** belief that the future of education will be shaped by **innovators** who blend empathy, vision, and courage with the best of AI's capacity for efficiency and adaptability. Every new platform, policy, or pilot is ultimately a reflection of those core values: you prioritize humanity, creativity, and authentic growth over outdated tasks or mere technology for technology's sake. AI enables you, and your kids, to bring their ideas and dreams to life like never before.

Remember:

1. **Human Connection Guides the Way**
   AI might handle routine lessons, but you are the mentor who sees each student's unique gifts, who coaches them through challenges, and who reminds them that technology serves human potential—not the other way around.

2. **Collaboration Multiplies Impact**
   Partner with colleagues, parents, community groups, and your students. Share wins, confront challenges together, and keep refining each initiative until it truly empowers everyone involved.

3. **Leadership Is a Mindset, Not a Title**
   No matter where you stand in the educational hierarchy—teacher, principal, parent volunteer—you can champion the responsible, mindful use of AI. You can create safe spaces for experimentation, cultivate trust, and celebrate achievements that inspire others.

4. **Take Bold Yet Thoughtful Action**
   Fear and inertia are real obstacles. Overcome them by focusing on pilot success stories and ongoing PD, seeking diverse funding avenues, and spotlighting genuine results—like student breakthroughs, teacher satisfaction, and creativity unleashed.

When the dust settles, it's you—**the educator, the leader**—who ensures that AI-driven reforms fulfill their promise of **personalization without dehumanization**, **efficiency without losing connection**, and **innovation without ethical shortcuts**. Your voice and vision matter more than ever. Embody that confidence, and you'll become a beacon of hope for colleagues and families navigating these new educational frontiers.

So carry this torch bravely. Blend the best of humanity with the promise of technology. Let your classroom (and your life's work) become a testament to the unstoppable combination of AI's adaptability and the ever-essential human spirit that guides and ignites real learning. So remove artificial limitations. Set your kids, and yourselves, up for success in this rapidly evolving era of AI by adapting and building the future you've been wanting to create. This is your moment.

Empowered by AI, driven by your passion for teaching—the future belongs to those who dare to reimagine what education can be.

REFERENCES

1. 21stCentEd. Building critical thinking skills with AI tools in the classroom. [Website]. https://21stcented.com/building-critical-thinking-skills-with-ai-tools-in-the-classroom/

2. ABCmouse. Retrieved from https://www.abcmouse.com

3. Adventure Academy. Retrieved from https://www.adventureacademy.com

4. AI for Education. AI in education 101 for parents: Cheat sheet 26. [Website]. https://www.aiforeducation.io/ai-resources/ai-in-education-101-for-parents

5. Ariely, D. (2008). Predictably irrational: The hidden forces that shape our decisions. HarperCollins.

6. Bloom, B. S. (1984). The 2 Sigma Problem: The search for methods of group instruction as effective as one-to-one tutoring. Educational Researcher, 13(6), 4–16.

7. BrainPOP Jr. Retrieved from https://jr.brainpop.com

8. Brookings. Digital tools for real-time data collection in education. [Website]. https://www.brookings.edu/articles/digital-tools-for-real-time-data-collection-in-education/

9. Campbell, R. AI gamification and growth mindset: A new era in education. [Website]. https://richardccampbell.com/game-based-learning-growth-mindset-a-new-era-in-education/

10. CCE Finland. The future of learning: How artificial intelligence and education merge to reshape our world. [Website]. https://www.ccefinland.org/post/the-future-of-learning-how-artificial-intelligence-and-education-merge-to-reshape-our-world

11. Clear, J. (2018). Atomic habits: An easy & proven way to build good habits & break bad ones. Avery.

12. Code.org. Retrieved from https://code.org

13. Common Sense Education. Classroom tools that use AI. [Website]. https://www.commonsense.org/education/lists/classroom-tools-that-use-ai

14. Cornell University. Ethical AI for teaching and learning. [Website]. https://teaching.cornell.edu/generative-artificial-intelligence/ethical-ai-teaching-and-learning

15. Csikszentmihalyi, M. (1990). Flow: The psychology of optimal experience. Harper & Row.

16. Culican, J. Embracing AI: The future of life skills education. [LinkedIn post]. https://www.linkedin.com/pulse/embracing-ai-future-life-skills-education-jamie-culican-j4loe

17. Digital Promise. (2024, June 18). AI literacy: A framework to understand, evaluate, and use emerging technology. [Website]. https://digitalpromise.org/2024/06/18/ai-literacy-a-framework-to-understand-evaluate-and-use-emerging-technology/

18. DigitalDefynd. (2025). Use of AI in schools [25 case studies] [2025]. [Website]. https://digitaldefynd.com/IQ/ai-in-schools-case-studies/

19. Disco. (2025). Top 5 AI tools for collaborative learning in 2025. [Website]. https://www.disco.co/blog/ai-tools-for-collaborative-learning

20. DreamBox. Retrieved from https://www.dreambox.com

21. Duolingo. Retrieved from https://www.duolingo.com

22. Dweck, C. S. (2006). Mindset: The new psychology of success. Random House.

23. Eastrise. Leveraging AI for financial literacy: A comprehensive guide. [Website]. https://www.eastrise.com/blog/leveraging-ai-for-financial-literacy-a-comprehensive-guide/

24. EdSurge. (2024, September 18). How AI can foster creative thinking in the classroom and beyond. [Website]. https://www.edsurge.com/news/2024-09-18-how-ai-can-foster-creative-thinking-in-the-classroom-and-beyond

25. EdTech Magazine. (2023, November). AI for teachers: Defeating burnout and boosting productivity. [Website]. https://edtechmagazine.com/k12/article/2023/11/ai-for-t

eachers-defeating-burnout-boosting-productivity-perfcon

26. EDU-AI. Empowering PBL with AI: Your co-teacher for the project-based learning journey. [Website]. https://edu-ai.org/empowering-pbl-with-ai-your-co-teacher-for-the-project-based-learning-journey/

27. EducationNext. (2023). Best AI tools for education and learning (2023). [Website]. https://www.educationnext.in/posts/best-ai-tools-for-education-and-learning-2023

28. eLearning Industry. (2025). AI in education: Personalized learning platforms in 2025. [Website]. https://elearningindustry.com/ai-in-education-personalized-learning-platforms

29. eLearning Industry. The science and benefits of gamification in eLearning. [Website]. https://elearningindustry.com/science-benefits-gamification-elearning

30. Epic. Retrieved from https://www.getepic.com

31. Fabulous. https://www.thefabulous.co

32. Forbes Technology Council. (2024, July 22). Personalized learning and AI: Revolutionizing education. Forbes. https://www.forbes.com/councils/forbestechcouncil/2024/07/22/personalized-learning-and-ai-revolutionizing-education/

33. Forest. https://www.forestapp.cc

34. Goodspeed Studio. AI in education: Best practices and strategies for implementing AI in the classroom. [Web-

site]. https://goodspeed.studio/blog/ai-in-education-best-practices-and-strategies-for-implementing-ai-in-the-classroom

35. Google. Google.org announces new AI funding for students and educators. [Website]. https://blog.google/outreach-initiatives/google-org/google-ai-initiatives-funding-educators-students/

36. Grammarly. https://www.grammarly.com

37. GUVI. 10 must-have AI tools for personalized learning. [Website]. https://www.guvi.com/blog/ai-tools-for-personalized-learning/

38. https://doi.org/10.3102/0013189X013006004

39. IBL. AI Mentor (K-12). [Website]. https://ibl.ai/product/ai-mentor-k-12

40. IGI Global. AI-enhanced education: Bridging educational disparities. [Website]. https://www.igi-global.com/chapter/ai-enhanced-education/345058

41. Impact of educational technology on teacher stress and well-being. PubMed Central. https://pmc.ncbi.nlm.nih.gov/articles/PMC7827099/

42. iReady. Retrieved from https://www.curriculumassociates.com/products/iready

43. Jasper. https://www.jasper.ai

44. Khan Academy Kids. Retrieved from https://learn.khanacadem

y.org/khan-academy-kids

45. Khan Academy. Retrieved from https://www.khanacademy.org

46. Kodexo Labs. AI in adaptive assessment: Benefits, tools and challenges. [Website]. https://kodexolabs.com/the-advantages-of-ai-in-adaptive-assessment/

47. Labster. https://www.labster.com

48. Merge EDU (Merge Cube). https://www.mergeedu.com

49. Meta-analysis of student engagement and its influencing factors. PubMed Central. https://pmc.ncbi.nlm.nih.gov/articles/PMC9855184/

50. MyFitnessPal. https://www.myfitnesspal.com

51. Newport, C. (2016). Deep work: Rules for focused success in a distracted world. Grand Central Publishing.

52. Notion. Notion AI. https://www.notion.so/product/ai

53. NWEA. (2020). 2020 MAP Growth normative data overview. Portland, OR: NWEA.

54. Panorama Education. A comprehensive guide to 21st century skills. [Website]. https://www.panoramaed.com/blog/comprehensive-guide-21st-century-skills

55. Pedagog.ai. Navigating screen time in the age of AI: A guide for educators. [Website]. https://pedagog.ai/navigating-screen-time-in-the-age-of-ai-a-guide-for-educators/

56. PhET Interactive Simulations. https://phet.colorado.edu

57. Photomath. Retrieved from https://photomath.com

58. Price, M. (2024). 2-Hour Learning White Paper. Alpha School.

59. Quillionz. Retrieved from https://www.quillionz.com

60. Quizizz. https://quizizz.com

61. Replit. Replit AI. https://replit.com/site/ai

62. Rethinking Schools. Alternatives to standardized tests. [Website]. https://rethinkingschools.org/articles/alternatives-to-standardized-tests/

63. Robbins, M. (2017). The 5 second rule: Transform your life, work, and confidence with everyday courage. Savio Republic.

64. Scratch. Retrieved from https://scratch.mit.edu

65. Sinek, S. (2009). Start with why: How great leaders inspire everyone to take action. Portfolio.

66. Skills Caravan. Personalised learning and adaptive learning algorithms are revolutionising education. [Website]. https://www.skillscaravan.com/post/personalised-learning-and-adaptive-learning-algorithms-are-revolutionising-education

67. SkoolOfCode. AI tools for kids to make learning fun and engaging. [Website]. https://skoolofcode.us/blog/ai-tools-for-kids-making-learning-fun-and-engaging/

68. Smythos. Real-world case studies of human-AI collaboration. [Website]. https://smythos.com/ai-agents/agent-architectures/human-ai-collaboration-case-studies/

69. Southern Regional Education Board. Overcoming resistance to AI in education initiatives. [Website]. https://www.sreb.org/blog-post/breaking-barriers-overcoming-resistance-ai-education-initiatives

70. Speeko AI. https://www.speeko.co

71. StoryLab AI. https://storylab.ai

72. Strides. https://www.stridesapp.com

73. Tavargere, Z. How I transitioned into an "AI engineer." [LinkedIn post]. https://www.linkedin.com/pulse/how-i-transitioned-ai-engineer-zahiruddin-tavargere-ivxyc

74. Teachers College, Columbia University. AI literacy for educators. [Website]. https://www.tc.columbia.edu/tcacademy/programs/all-offerings/ai-literacy-for-educators/

75. Teachflow.ai. AI in education: Global success stories unveiled. [Website]. https://teachflow.ai/ai-in-education-global-success-stories-unveiled/

76. Walden University. Ten myths about artificial intelligence in education. Walden University. https://scholarworks.waldenu.edu/cgi/viewcontent.cgi?article=15086&context=hlrc

77. IEEE Xplore. The impact of AI-driven educational simulations

on... IEEE Xplore. https://ieeexplore.ieee.org/document/10427299/

78. The University of Texas at San Antonio. The growing demand for AI skills and expertise in the job market. [Website]. https://www.utsa.edu/pace/news/growing-demand-for-ai-skills-and-expertise-in-the-job-market.html

79. U.S. Department of Education. Artificial intelligence and the future of teaching and learning. [PDF]. https://www.ed.gov/sites/ed/files/documents/ai-report/ai-report.pdf

80. UNESCO. Artificial intelligence in education. [Website]. https://www.unesco.org/en/digital-education/artificial-intelligence

81. UNESCO. Use of AI in education: Deciding on the future we want. [Website]. https://www.unesco.org/en/articles/use-ai-education-deciding-future-we-want

82. University of Pittsburgh. Reconsidering education policy in the era of generative AI. [PDF]. https://www.research.pitt.edu/sites/default/files/reconsidering_education_policy_in_the_era_of_generative_ai.pdf

83. University of San Diego. 39 examples of artificial intelligence in education. [Website]. https://onlinedegrees.sandiego.edu/artificial-intelligence-education/

84. Woebot. https://woebothealth.com

85. World Schools. Challenges of traditional schooling. [Website]. https://world-schools.com/challenges-of-traditional-schooling/

86. Wysa. Retrieved from https://www.wysa.io

87. Yoodli. Retrieved from https://app.yoodli.ai

88. DonorsChoose. https://www.donorschoose.org

89. Grants.gov. https://www.grants.gov

90. ISTE (International Society for Technology in Education). Professional Development & AI Education. https://www.iste.org/areas-of-focus/ai-education

Appendix

—Curated AI & Enrichment Tool List For the Curious Educator

1. AI Tutors

- **Khan Academy AI**
    - **What It Does**: Uses AI-powered feedback to guide students through math, science, humanities, and more.
    - **Enables Kids To**: Get instant help, personalized hints, and tailored practice recommendations.
- **DreamBox**
    - **What It Does**: Adaptive math learning platform offering real-time evaluation and auto-adjusted lesson paths.
    - **Enables Kids To**: Progress at their own pace, filling in knowledge gaps and accelerating when ready.
- **Photomath**

  - **What It Does**: Allows users to take a picture of a math problem and see step-by-step solutions.
  - **Enables Kids To**: Understand the full process of solving math problems, not just the answer.

- **iReady**
  - **What It Does**: Diagnostic-driven instruction in math and reading; combines assessments and personalized lessons.
  - **Enables Kids To**: Target specific areas for improvement, track progress, and build strong foundational skills.

2. AI Writing

- **Grammarly AI**
  - **What It Does**: Provides grammar, spelling, and style corrections in real time.
  - **Enables Kids To**: Develop better writing habits, refine essays, emails, and creative writing pieces.
- **Quillionz**
  - **What It Does**: Uses AI to generate reading comprehension questions from any text.
  - **Enables Kids To**: Practice critical thinking, quiz themselves, and improve reading comprehension.

- **Jasper AI (Writing & Public Speaking)**
  - **What It Does**: Generates text and speaking prompts; can also suggest outlines and ideas.
  - **Enables Kids To**: Brainstorm effectively, develop public-speaking scripts, and polish writing projects.

3. AI Coding

- **Replit AI**
  - **What It Does**: Online coding environment with AI assistance, offering real-time code suggestions.
  - **Enables Kids To**: Learn multiple programming languages collaboratively, debug quickly, and build projects with AI guidance.
- **Code.org**
  - **What It Does**: Interactive lessons teaching kids basic programming concepts with block-based and text-based activities.
  - **Enables Kids To**: Progress from beginner-friendly coding tutorials to more advanced topics, encouraging creativity.
- **Scratch AI**
  - **What It Does**: Visual drag-and-drop programming environment with AI-inspired blocks and extensions.

- **Enables Kids To**: Create interactive stories, animations, and games while exploring foundational AI concepts.

4. AI Creativity

- **Ideogram & DALL·E**
  - **What It Does**: AI model that generates images from text prompts.
  - **Enables Kids To**: Experiment with digital art, explore visual storytelling, and expand creative expression.
- **Runway ML**
  - **What It Does**: Provides a suite of AI-powered tools for video editing, image generation, and other creative tasks.
  - **Enables Kids To**: Produce multimedia projects, learn advanced editing techniques, and explore innovative visual effects.
- **StoryLab AI**
  - **What It Does**: AI tool for generating story plots, character outlines, and creative writing prompts.
  - **Enables Kids To**: Jump-start fiction writing, enhance narrative structure, and develop storytelling skills.
- Gamma AI

    - **What It Does**: AI tool that generates compelling presentations that bring ideas to life.
    - **Enables Kids To**: Put together amazing presentation decks to showcase their ideas.

5. AI Public Speaking

- **Yoodli AI**
    - **What It Does**: Analyzes recorded speeches or presentations to offer feedback on pacing, filler words, and clarity.
    - **Enables Kids To**: Gain confidence, refine delivery, and track progress in real-time speech performance.
- **Speeko AI**
    - **What It Does**: Uses speech recognition to assess clarity, intonation, and word choice.
    - **Enables Kids To**: Practice in a safe environment, receive guided practice exercises, and build public-speaking skills.
- **Jasper AI (Speech)**
    - **What It Does**: Creates draft scripts for speeches or presentations and provides suggestions for improvements.
    - **Enables Kids To**: Polish speaking engagements, structure logical arguments, and practice tone and pacing.

6. AI for Parents

- **Google Read Along**
  - **What It Does**: Uses speech recognition to guide children through reading activities, offering corrective feedback.
  - **Enables Kids To**: Improve reading fluency and pronunciation while parents can monitor progress.
- **Duolingo AI**
  - **What It Does**: Employs AI to tailor language-learning lessons; covers a wide array of languages.
  - **Enables Kids To**: Practice language skills in bite-sized gamified lessons and stay motivated via progress tracking.
- **Wysa AI**
  - **What It Does**: AI-driven "chat buddy" for mental health check-ins and emotional well-being.
  - **Enables Kids To**: Build resilience, share feelings, and develop coping strategies with supportive prompts.

7. Comprehensive E-Learning Platforms

- **BrainPOP & BrainPOP Jr.**
  - **What It Does**: Offers short, animated videos on science, social studies, math, and more for K–8.

    - **Enables Kids To**: Engage visually, build early content knowledge, and follow up with interactive quizzes.

- **Khan Academy (Main Site)**
    - **What It Does**: Free lessons in math, science, humanities, test prep, and more for all ages.
    - **Enables Kids To**: Access an entire library of self-paced courses and practice exercises with detailed feedback.

- **Khan Academy Kids**
    - **What It Does**: Free program for ages 2–7 focusing on early literacy, reading, writing, language, and math.
    - **Enables Kids To**: Develop foundational reading and math skills, stay engaged with interactive characters, and track progress.

- **IXL**
    - **What It Does**: Offers a range of academic practice exercises in math, language arts, science, and social studies.
    - **Enables Kids To**: Receive immediate feedback, track skill mastery, and practice targeted areas by grade level.

- **Adventure Academy**
    - **What It Does**: Immersive online environment covering math, language arts, science, and social studies (ages 8–13).

  - **Enables Kids To**: Explore virtual worlds, complete educational quests, and engage in gamified learning.

- **ABCmouse**
  - **What It Does**: Online curriculum (ages 2–8) covering language arts, math, science, social studies, and art through gamification.
  - **Enables Kids To**: Follow a structured learning path, earn rewards, and gain familiarity with core subjects at an early age.
- **Prodigy**
  - **What It Does**: Fantasy-themed math platform (ages 6 and up) with adaptive difficulty and battle-based gameplay.
  - **Enables Kids To**: Stay motivated through adventure-style play, earn rewards, and improve core math skills.
- **PBS KIDS / PBS Parents Play and Learn**
  - **What It Does**: Provides a wide variety of games, videos, and parent-child activities aligned with PBS KIDS TV programs.
  - **Enables Kids To**: Engage with familiar characters, strengthen parent-child interaction, and explore STEM, reading, and social-emotional skills.
- **Epic**
  - **What It Does**: Digital library with 40,000+ e-books for kids up to age 12.

  - **Enables Kids To**: Develop a love of reading, access an extensive collection anytime, and track reading progress.

- **Starfall**
  - **What It Does**: Focuses on early literacy and math (Pre-K–3), featuring phonics games and reading exercises.
  - **Enables Kids To**: Learn foundational reading skills, practice spelling, and build confidence in basic math.

- **National Geographic Kids**
  - **What It Does**: Interactive platform (ages 4–12) covering animals, geography, and environmental science.
  - **Enables Kids To**: Explore the natural world via videos, quizzes, and articles—fueling curiosity.

- **Teach Engineering**
  - **What It Does**: Offers hands-on STEM activities for K–12, from simple projects to more advanced engineering tasks.
  - **Enables Kids To**: Apply math/science in real-world contexts, build problem-solving skills, and experiment with design thinking.

- **NASA Kids' Club**
  - **What It Does**: Collection of space-themed games and activities (ages 5–8) from NASA.

    - **Enables Kids To**: Discover basic astronomy, aerospace concepts, and spark curiosity about the universe.

- **Ted-Ed**
    - **What It Does**: Curated educational videos and lessons on various topics, from history to science.
    - **Enables Kids To**: Explore engaging mini-lectures, take interactive quizzes, and deepen understanding through rich animations.

8. Coding & STEM Exploration

- **Blockly (MakeWonder.com/apps/blockly)**
    - **What It Does**: Visual programming interface that pairs with Wonder Workshop robots (like Dash & Dot).
    - **Enables Kids To**: Learn coding logic, problem-solving, and robotics in a hands-on, fun environment.
- **Scratch Jr.**
    - **What It Does**: Simplified version of Scratch for younger children (ages 5–7), teaching block-based coding basics.
    - **Enables Kids To**: Create animations, learn sequencing, and practice computational thinking.

9. Music

- **Yousician**
    - **What It Does**: Interactive tool teaching piano, guitar, bass, or ukulele via real-time feedback on playing.
    - **Enables Kids To**: Develop musical skills in a fun, game-like environment with auto-assessment and progress tracking.

How to Use This Appendix

- **Finding the Right Tool**: Decide which subject or skill is a priority (e.g., math remediation, early reading, AI coding, or creative arts).
- **Combining Apps**: Mix and match to cover multiple skill areas —e.g., IXL for math and reading, and Blockly for coding.
- **Parental Involvement**: Tools labeled "AI for Parents" or "PBS Parents" can help families guide kids' learning journeys with data insights, mental health support, or co-play activities.
- **Monitoring Progress**: Many of these apps include dashboards or progress trackers, so parents and educators can watch how kids evolve over time.

This curated list showcases an array of modern educational and AI-powered platforms, each designed to engage children at their level, provide personalized feedback, or spark new interests to help them discover their passions.

Made in the USA
Monee, IL
04 November 2026

67378722R00075